The Natural History of

SHREWS

The Natural History of Mammals Series
Comstock Publishing Associates
a division of Cornell University Press

Published in the United Kingdom as
Christopher Helm Mammal Series
Edited by Dr Ernest Neal, MBE, former President of the
Mammal Society

The Natural History of Deer
Rory Putman

The Natural History of Weasels and Stoats
Carolyn M. King

The Natural History of Shrews
Sara Churchfield

Forthcoming titles include:

The Natural History of the Wild Cats
Andrew Kitchener

The Natural History of

SHREWS

Sara Churchfield

COMSTOCK PUBLISHING ASSOCIATES
A division of CORNELL UNIVERSITY PRESS
Ithaca, New York

First published 1990 by Christopher Helm Ltd,
a subsidiary of A & C Black, London and Cornell University Press.

Library of Congress Cataloging-in-Publication Data

    Churchfield, Sara.
      The natural history of shrews / Sara Churchfield : [line drawings by Sally Alexander and Sara Churchfield].
         p.   cm. — (The Natural history of mammals series)
      Includes bibliographical references (p.   ) and index.
      ISBN 0-8014-2595-6 (alk. paper)
      1. Shrews.   I. Title.   II. Series.
    QL737.I56C48   1990
    599.3'3—dc20

Photoset by Florencetype Ltd, Kewstoke, Avon
Printed and bound in Great Britain by Hartnolls Ltd, Bodmin, Cornwall

Contents

Now go thy ways; thou has tamed a curst shrew.
'Tis a wonder, by your leave, she will be tamed so.

Black-and-white plates

Figures

Tables

Acknowledgements

I am grateful to Ernest Neal for giving me the opportunity to write a book on shrews and for encouraging me throughout its production. Thanks to Robert Kirk, Alan Marshall and the staff at Christopher Helm for their friendly and efficient service in the production of this book, and to Sally Alexander for her hard work in the preparation of the line drawings. I am indebted to David Hosking for providing such excellent photographs.

Special thanks go to John Gurnell and Gordon Kirkland for their valuable comments and suggestions on the manuscript, and also to Sarah George and Ernest Neal for their advice. I am grateful to Rainer Hutterer, Cornelis Neet and Gordon Kirkland for allowing me to make use of figures and tables. I thank all of them, together with Michel Genoud, Jacques Hausser, John Griffith, Jo Merritt, Zdzislaw Pucek and Peter Vogel for sharing and encouraging my interest in shrews. I happily acknowledge the help of John Hollier for his good-natured assistance in the field, John Henry who pushed, cajoled and entertained me through this production, and the many other colleagues, students and friends, too numerous to mention by name, who have aided and inspired me.

My warmest thanks go to my mother and father who have suffered, encouraged and supported me and my menagerie over the years.

Series Editor's foreword

In recent years there has been a great upsurge of interest in wildlife and a deepening concern for nature conservation. For many there is a compelling urge to counterbalance some of the artificiality of present-day living with a more intimate involvement with the natural world. More people are coming to realise that we are all part of nature, not apart from it. There seems to be a greater desire to understand its complexities and appreciate its beauty.

This appreciation of wildlife and wild places has been greatly stimulated by the world-wide impact of natural-history television programmes. These have brought into our homes the sights and sounds both of our own countryside and of far-off places that arouse our interest and delight.

In parallel with this growth of interest there has been a great expansion of knowledge and, above all, understanding of the natural world — an understanding vital to any conservation measures than can be taken to safeguard it. More and more field workers have carried out painstaking studies of many species, analysing their intricate behaviour, relationships and the part they play in the general ecology of their habitats. To the time-honoured techniques of field observations and experimentation has been added the sophistication of radio-telemetry whereby individual animals can be followed, even in the dark and over long periods, and their activities recorded. Infra-red cameras and light-intensifying binoculars now add a new dimension to the study of nocturnal animals. Through such devices great advances have been made.

This series of volumes aims to bring this information together in an exciting and readable form so that all who are interested in wildlife may benefit from such a synthesis. Many of the titles in the series concern groups of related species such as otters, squirrels and rabbits so that readers from many parts of the world may learn about their own more familiar animals in a much wider context. Inevitably more emphasis will be given to particular species withn a group as some have been more extensively studied than others. Authors too have their own special interests and experience and a text gains much in authority and vividness when there has been personal involvement.

Many natural history books have been published in recent years which have delighted the eye and fired the imagination. This is wholly good.

But it is the intention of this series to take this a step further by exploring the subject in greater depth and by making available the results of recent research. In this way it is hoped to satisfy to some extent at least the curiosity and desire to know more which is such an encouraging characteristic of the keen naturalist of today.

Ernest Neal
Bedford

1 Characteristics of shrews

GENERAL FEATURES

Shrews are small, short-legged, mouse-like mammals with long, pointed snouts and short, dense fur, usually dark brown in colour. They are predatory, feeding mainly on small invertebrates. Their eyes are extremely small and almost hidden amongst the fur. The ears are small and rounded and often quite inconspicuous. Most shrews have long, hairy tails.

Shrews are placed in the order Insectivora, as are their relatives the moles, hedgehogs, otter shrews and tenrecs which they resemble in many ways. Along with their fellow insectivores, shrews number amongst the most ancient of mammals. They probably first evolved soon after the dinosaurs disappeared, in the late Eocene/early Oligocene epoch some 38 million years ago (Findley, 1967; Repenning, 1967; Yates, 1984), and they have remained virtually unchanged since then, although they have undergone a slight reduction in size.

These insectivores show some of the earliest mammalian features and retain a number of primitive characters. They have a generalised and unspecialised body plan with a simple plantigrade mode of locomotion whereby they run along with the sole of the foot placed flat on the ground. Each foot possesses five toes or digits, each terminating in a simple claw. Their skulls are long and narrow and the brain is small. The zygomatic arches on the lateral sides of the skull so characteristic of many mammals, including rodents, are lacking in shrews, and the mandible possesses a double articulating surface which is also reminiscent of the more primitive condition. The cerebral hemispheres, which occupy the largest part of the brain in more advanced mammals, are rather small in shrews, suggesting that intelligence and manipulative abilities in these tiny mammals are not great. But the olfactory lobes of the brain are relatively large and well developed, reflecting the importance of the sense of smell in their daily lives. The skulls of a shrew and a mouse are compared in Figure 1.1.

In most shrews, the genital and urinary systems have a common opening via a cloaca, which is reminiscent of the primitive, reptilian condition. In more modern mammals, including ourselves, the openings

1

Figure 1.1 The skull of a shrew (insectivore) and a mouse (rodent) compared

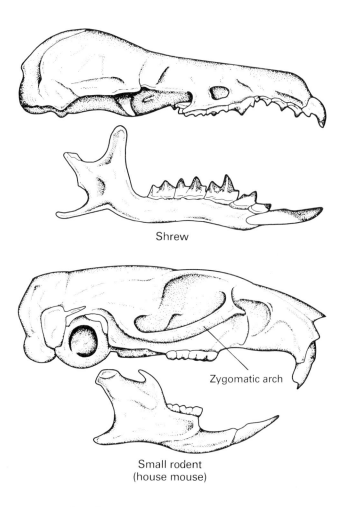

Shrew

Zygomatic arch

Small rodent
(house mouse)

are quite separate. Another primitive feature of shrews is the position of the males' testes: these are retained within the abdominal cavity and do not descend into a sac or scrotum outside the body as in the majority of mammals.

The teeth of shrews, on the other hand, are surprisingly specialised and very characteristic. The incisors are particularly distinctive, being large and prominent, as Figure 1.2 shows. The first upper incisors are not only enlarged but they each possess two hooked cusps which project downwards plus a large cusp at the base. The first lower incisors, too, are large but also slightly procumbent, pointing forwards and upwards. The smaller cheek teeth possess single, pointed cusps while the larger molar teeth have sharp cusps arranged in a W-pattern. The upper and lower jaws are articulated in such a way as to allow little rotary chewing motion but considerable fore and aft movement of the lower jaw. This may allow the incisor teeth to act rather like forceps or pincers as prey

*Figure 1.2 The anterior teeth of a red-toothed shrew (*Sorex araneus*) and a white-toothed shrew (*Crocidura russula*)*

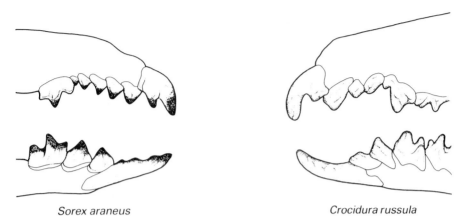

Sorex araneus   Crocidura russula

are picked up, the food then being passed to the cheek teeth whose sharp, pointed cusps can pierce and fragment the hard, chitinous exoskeletons of beetles and other such prey.

Like other mammals, shrews grow both a milk and a permanent dentition, but the first or milk teeth are shed as an embryo so that young shrews are born with a permanent set of 26–32 teeth (depending on the species) already in place. These teeth do not grow as they wear away, as in the case of rodents, but must last a lifetime.

THE DIVERSITY OF SHREWS

The family Soricidae, to which shrews belong, contains a multitude of species occupying various ecological niches and modes of life. The different genera of living shrews, their common names, and the number of species they comprise are shown in Table 1.1. Currently, some 266 species are recognised, belonging to 20 different genera. They are something of a taxonomist's nightmare since many species are so similar that they are distinguishable only through detailed study of their skulls and teeth or chromosome morphology. Some species, including members of the genera *Solisorex*, *Feroculus*, *Nectogale* and *Megasorex*, are apparently very rare, and few specimens have ever been caught. Their habits are virtually unknown.

Shrews range in size from tiny creatures smaller than a mouse to specimens as large as a rat. One of the smallest is the Etruscan shrew, *Suncus etruscus*, found in southern Europe and parts of Africa and Asia, which can weigh a mere 2g as an adult with a head and body length of only 35mm. Similarly minute species include *Suncus remyi*, *Sorex nanus* and *Sorex minutissimus*. Not only are these the smallest of shrews but also amongst the smallest of all mammals alive today. The largest shrew is probably *Suncus murinus* which inhabits parts of tropical Africa and Asia and whose weight can reach 106g or more with a head and body length of some 150mm. *Crocidura odorata* and *C. occidentalis*, found in

Table 1.1   The diversity of shrews: the genera and numbers of species

Genus	Common name	No. of species
Anourosorex	Mole shrew	1
Blarina	American short-tailed shrews	4
Blarinella	Asiatic short-tailed shrew	1
Chimarrogale	Oriental water shrews	4
Crocidura	White-toothed shrews	125
Cryptotis	Small-eared shrews	12
Diplomesodon	Turkestan desert shrew/Piebald shrew	1
Feroculus	Kelaart's long-clawed shrew	1
Megasorex	Mexican shrew	1
Myosorex	Mouse shrews; forest shrews	12
Nectogale	Tibetan web-footed water shrew	1
Neomys	Eurasian water shrews	3
Notiosorex	American desert shrew	1
Paracrocidura	Cameroon shrew	2
Scutisorex	Armoured shrew; Hero shrew	1
Solisorex	Pearson's long-clawed shrew	1
Sorex	Red-toothed or Long-tailed shrews	64
Soriculus	Asiatic mountain shrews	10
Suncus	Musk shrews	14
Sylvisorex	Forest musk shrews	7
Total		266

the tropical forests of West Africa, are also giants amongst shrews, but few specimens of these species have ever been caught. Most shrews, however, are small or medium in size, such as the common shrew (*Sorex araneus*) with a head and body length of 48–80mm, a tail length of 24–44mm and a weight of 5–14g.

DISTRIBUTION AND HABITATS

Shrews are found throughout Africa, Asia, Europe and North and Central America (see Figure 1.3). In fact they occur on all major land areas except Australia, Tasmania, New Zealand, Antarctica, Greenland, Iceland, the Arctic islands, Ungava, the West Indies and some of the Pacific islands. In South America they are found only in the north-west corner where they are represented by several species of *Cryptotis*. Not only is *Cryptotis* the only shrew present in South America, it is the only insectivore to penetrate south into the Neotropics. These shrews are thought to have reached South America via a land bridge from North America during the Pliocene epoch. An indication of the distribution and diversity of shrews is given in Table 1.2.

Although shrews mostly occupy moist, terrestrial environments where there is an abundance of vegetation cover and a wealth of invertebrate prey, they are extremely widespread and can be found in many different terrestrial and even some aquatic habitats. They are found in woodlands, forests, grasslands, hedgerows and scrublands of all kinds in both

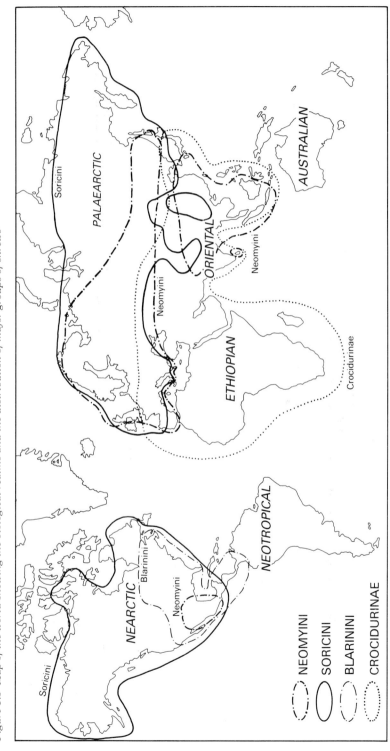

Figure 1.3 Map of the world showing the biological realms and the distribution of major groups of shrews

Table 1.2 The distribution and diversity of shrews: the numbers of shrew species of each genus occurring in different geographical regions (based on data from Corbet and Hill, 1980; Novak and Paradiso, 1983; and Hutterer, 1985)

GENUS	North America	Central/ South America	Europe	Africa	Central/ North Asia	Asia Minor	China & Himalayas	India Nepal & Sri Lanka	S.E. Asia & islands	Total species
Sorex	27	12	14		12		13		1	79
Soriculus							7	1	2	10
Blarina	4									4
Blarinella							1			1
Cryptotis	1	12								13
Notiosorex	1	1								2
Megasorex		1								1
Crocidura			3	83	3	5	5	2	29	130
Suncus			1	6	1			3	8	19
Feroculus								1		1
Solisorex								1		1
Paracrocidura				2						2
Sylvisorex				7						7
Myosorex				12						12
Diplomesodon					1					1
Scutisorex				1						1
Anourosorex							1		1	2
Neomys			2		2					4
Chimarrogale							3		2	5
Nectogale							1			1
Total Species	33	26	20	111	19	5	31	8	43	

temperate and tropical regions. The water shrews, including the European water shrew *Neomys fodiens* and the American water shrew *Sorex palustris*, are found beside streams, ponds, drainage ditches and even icy mountain torrents, in which they forage for aquatic prey. At the other extreme there are species such as *Notiosorex* and *Diplomesodon pulchellum* which commonly occur in arid regions where they are found in semi-desert scrublands and sandy habitats. The Scilly shrew, *Crocidura suaveolens*, frequently occurs in littoral habitats: on the islands it occupies it may be found foraging amongst boulders and decaying seaweed on the strand line of the seashore or around marram grass on sand dunes.

Some shrews habitually live around human dwellings, and these include the large musk shrew *Suncus murinus* in the tropics and the so-called house shrew, *Crocidura russula*, familiar to householders in central and southern Europe. Even the common shrew, *Sorex araneus*, is often to be found in gardens around compost heaps and hedges. It has also been caught inside houses, having apparently arrived of its own accord and not by courtesy of the domestic cat.

Altitude presents no barrier to shrews, provided food and cover are available. The common shrew, *Sorex araneus*, has been caught amongst scree on a Scottish mountainside at 1,000m elevation. The Szechwan burrowing shrew, *Anourosorex squamipes*, occurs in mountain forests up to 3,100m, and in tropical regions shrews' ranges extend up to at least 3,760m on Mount Kilimanjaro.

Shrews are robust and hardy and have evolved to withstand a wide range of temperatures and other environmental extremes. It can be seen from their geographical distribution that they occupy habitats as diverse as freezing arctic tundra to steamy equatorial rain forest. While certain species or groups of species are typical of different latitudes and climatic regimes, some are able to exist in many habitat types under a great range of environmental conditions. Both the common shrew (*Sorex araneus*) and the pygmy shrew (*S. minutus*), for example, extend from the southernmost tips of Italy and Greece to the far north of Scandinavia at latitude 70°N.

Most shrews can climb and burrow a little but they generally live on the ground surface amongst vegetation, hiding up in nests in grass tussocks or under logs. They also occupy and modify underground tunnels and crevices created by other small mammals more adept at burrowing.

CLASSIFICATION

The classification of shrews and some of their close relatives in the order Insectivora is summarised in Figure 1.4. The family Soricidae is divided into two subfamilies, the Soricinae and the Crocidurinae. The Soricinae contains the red-toothed shrews, so-called because of the pigmentation of the tips of the teeth which ranges from yellow to dark purple but is commonly reddish-brown. This colouring is caused by the deposition of iron in the outer layer of enamel of the teeth (Dötsch and Koenigswald, 1978) which may increase their resistance to wear, a considerable advantage as the teeth do not grow through life. These red-toothed

Figure 1.4 Classification of living shrews

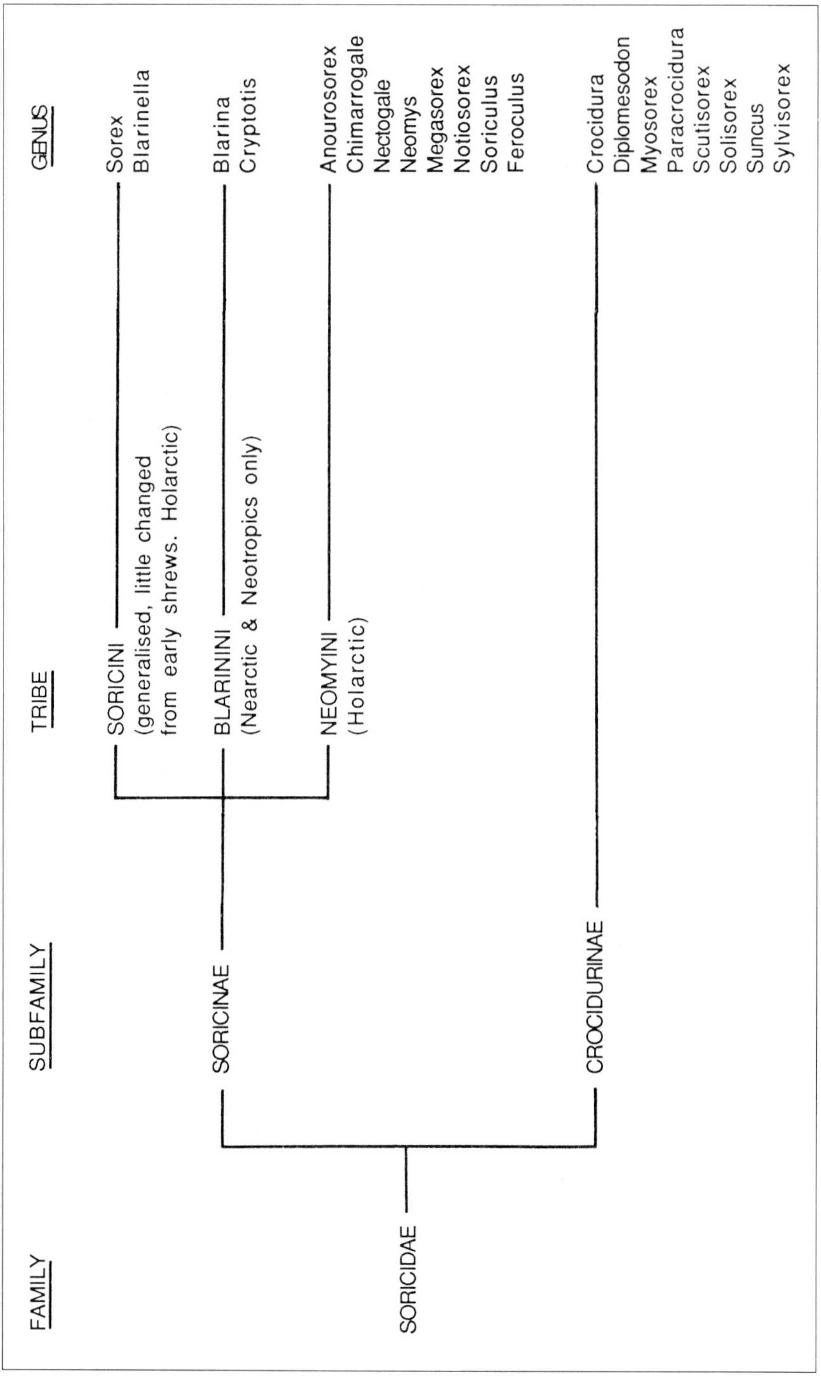

FAMILY	SUBFAMILY	TRIBE	GENUS
SORICIDAE	SORICINAE	SORICINI (generalised, little changed from early shrews. Holarctic)	Sorex, Blarinella
		BLARININI (Nearctic & Neotropics only)	Blarina, Cryptotis
		NEOMYINI (Holarctic)	Anourosorex, Chimarrogale, Nectogale, Neomys, Megasorex, Notiosorex, Soriculus, Feroculus
	CROCIDURINAE		Crocidura, Diplomesodon, Myosorex, Paracrocidura, Scutisorex, Solisorex, Suncus, Sylvisorex

shrews include many of the familiar northern temperate species of the genus *Sorex* such as the common and pygmy shrews (*Sorex araneus* and *S. minutus*), the masked shrew (*Sorex cinereus*), and the short-tailed shrew (*Blarina brevicauda*). This subfamily of shrews has a Holarctic and northern Neotropical distribution.

The Soricinae, in turn, is divided into three different tribes: the Soricini, the Neomyini and the Blarinini, since the various component genera form natural groups showing similarities in anatomy, physiology, distribution and evolutionary history.

The subfamily Crocidurinae includes the white-toothed shrews which lack the pigmentation in the teeth which is so characteristic of the red-toothed shrews. It comprises a large number of shrews, many of which belong to the genus *Crocidura* which includes species such as the Scilly shrew (*C. suaveolens*) and the house shrew (*C. russula*), both quite common in Europe. Although several species are Palaearctic in distribution, extending northwards into Europe, they are really tropical species centred on Africa and Asia where they are widespread and ubiquitous.

By reference to Tables 1.1 and 1.2, where the numbers of species belonging to each genus are shown, it can be seen that the three largest genera in terms of the numbers of species are *Crocidura* with some 125 species, *Sorex* with 64 species and *Suncus* with 14 species. These three genera are also geographically the most widespread.

EVOLUTIONARY HISTORY OF SHREWS

Since they are both ancient and diverse mammals, with an almost world-wide distribution, it is interesting to speculate on the evolution of shrews. Because of their small size, however, their fossil record is pretty scanty and the evolutionary relationships within this large family are poorly understood. The earliest fossils found which can be attributed to the family Soricidae do not appear until the middle Oligocene, and Repenning (1967) suggests that they probably originated in the late Eocene, some 30–40 million years ago. Five ancient subfamilies of shrews were recognised by Repenning, but more recent research by Reumer (1987) suggests there were only four subfamilies. Only two of these subfamilies are still extant. As we have seen, these are the Soricinae with a Holarctic and northern Neotropical distribution, and the Crocidurinae with a Palaearctic, Oriental and Ethiopian distribution (see Figure 1.3).

The earliest known fossil shrew is *Crocidosorex piveteaui* from the late Oligocene and early Miocene in Europe. It is thought to have been very similar to modern shrews, and is considered to be ancestral to the Soricidae. Repenning (1967) suggested that the Crocidurinae did not descend from this shrew, but may have diverged from ancestral stock, prior to the appearance of *Crocidosorex*, from a common ancestor as yet unknown. Reumer (1987), on the other hand, suggested that an ancient subfamily, the Crocidosoricinae, is ancestral to both the Crocidurinae and the Soricinae. The Crocidurinae appear in Europe in the early Miocene (Reumer, 1984).

Although the fossil record for shrews is far from complete, it does contain enough information now to propose a hypothesis for their phylogeny by comparing knowledge about existing fossils with new biochemical techniques which look for genetic similarities in extant species. It is possible to calculate divergence times of many species and groups of species by linking estimates of genetic distances to a protein molecular clock (see Thorpe, 1982). While this is possible for many mammals, it has proved a problem for shrews because there appear to be varying rates of protein change in different lineages.

Dr Sarah George has combined a knowledge of fossil data with the biochemical techniques of allozyme electrophoresis which use tissues from living species of shrew. Samples of tissues such as heart, liver and kidney from shrews have been subjected to starch-gel electrophoresis. Allelic frequencies, heterozygosity and the percentage of polymorphic loci were calculated for each species and used to estimate the phylogenetic similarities between species and groups of species. The fossil record was used to set dates for divergence points elucidated in the allozyme analyses.

By comparing evidence from different sources and different techniques in this way it is possible to establish a clearer view of the evolutionary relationships of soricine shrews, albeit still an interrupted one. Having first evolved in Europe, *Crocidosorex*, or its descendents, probably spread to the Nearctic during the Miocene via the area of the Bering Strait which was above sea level at this time. There it gave rise to *Antesorex* which is known from the late early Miocene of North America. Two descendents of *Antesorex* are then known from the late Miocene in the Nearctic, namely *Adeloblarina* and *Alluvisorex* (Repenning, 1967; Savage and Russell, 1983). *Adeloblarina* is considered to be ancestral to the tribe Blarinini whereas *Alluvisorex* is probably ancestral to the tribe Soricini (Repenning, 1967; Gureev, 1971). This was supported by George's (1986) allozyme data which suggested that, although there was a relatively low phenetic similarity between living representatives of these two tribes, they probably had a common ancestor in *Antesorex*. Subsequent evolution of the Blarinini occurred mainly in the Nearctic (Repenning, 1967; Choate, 1970; Jones *et al.*, 1984). The two genera within this tribe which are still alive today, namely *Blarina* and *Cryptotis*, probably diverged in the Pliocene (Repenning, 1967; George, 1986).

Alluvisorex, which was ancestral to the tribe Soricini, gave rise to two lines. The first spread from the Nearctic back to the Palaearctic in the early Pliocene, giving rise to the genus *Blarinella* (Repenning, 1967; Gureev, 1971). The second gave rise to *Sorex* which is a diverse genus with a wide distribution in both the Nearctic and the Palaearctic.

The third tribe of shrews, the Neomyini, containing the water shrews and several other similar types (see Figure 1.4), has an extremely poor fossil record and the relationships between the living genera have yet to be fully elaborated. Repenning (1967) proposed that this tribe of shrews diverged from the ancestral stock of the Soricinae some time in the early Miocene, probably in the Palaearctic since shrews belonging to this tribe are still distributed mainly throughout that region. The earliest known member of this tribe is *Amblycoptus* from the early Pliocene of Europe which then gave rise to two distinct lineages of Neomyini, one in the Palaearctic and one, probably more recently, in the Nearctic

(George, 1986). The earliest known neomyine fossil in North America is *Hesperosorex* from the late middle Pliocene. But, there are, as yet, no known intermediate forms between *Amblycoptus* and the ancestral *Crocidosorex*.

A summary of these lineages can be seen in Figure 1.5. There is still much to learn about the origins and phylogenetic relationships of shrews. While studies such as these may appear to be purely academic, they do in fact have considerable relevance to the characteristics of different types of shrew, their geographical distributions, and their adaptations and tolerances to a wide range of environmental conditions. As we shall see in subsequent chapters, shrews vary enormously with respect to their ecology, their behaviour and, particularly, their physiology, which has contributed greatly to their success in a wide range of climates.

ANATOMICAL AND MORPHOLOGICAL ADAPTATIONS OF SHREWS

Although shrews all share the same basic body plan and characteristic looks which makes them easy to distinguish from other small mammals, on closer inspection they reveal a great diversity of forms and adaptations to fill a variety of ecological niches and modes of life. Thus shrews have specialisations superimposed upon their generalised and rather primitive body plan. They have expanded to occupy terrestrial, subterranean, scansorial (climbing), semi-aquatic and psammophilic (sand-dwelling) modes of life, and show corresponding adaptations through some remarkable examples of adaptive radiation and convergent evolution. Many species are not restricted to a single mode of life. For instance water shrews such as *Neomys fodiens* will forage both on land and in freshwater. The short-tailed shrew *Blarina brevicauda*, besides being the most fossorial of North American shrews, is also a competent climber. An individual of this species was caught 1.9m up a tree where it had been attracted to bait in a trap. The common shrew (*Sorex araneus*) although primarily a dweller of the ground surface, has been found occupying the nests of harvest mice (*Micromys minutus*) in bushes.

The Skeleton: Feet, Tails and Vertebrae

The basic skeletal plan of shrews is exemplified by *S. araneus* in Figure 1.6 which has short legs, simple feet and a tail which is just over half the length of the head and body. But there are many examples of shrews which deviate slightly from this general plan and have become adapted for particular modes of life. *Anourosorex*, a mole-like, semi-fossorial shrew, has a mere stump of a tail about 10mm in length, and *Diplomesodon pulchellum*, the desert shrew, also has a much shortened tail, and particularly long claws to assist digging. Other supposed burrowing forms such as the rare *Feroculus* and *Solisorex* from Sri Lanka also possess particularly long claws on the fore feet. *Sylvisorex megalura* and *Soriculus leucops*, which are scansorial species, have much elongated tails which probably assists balance when clambering amongst bushes and rocks. These species also have rather elongated feet with relatively

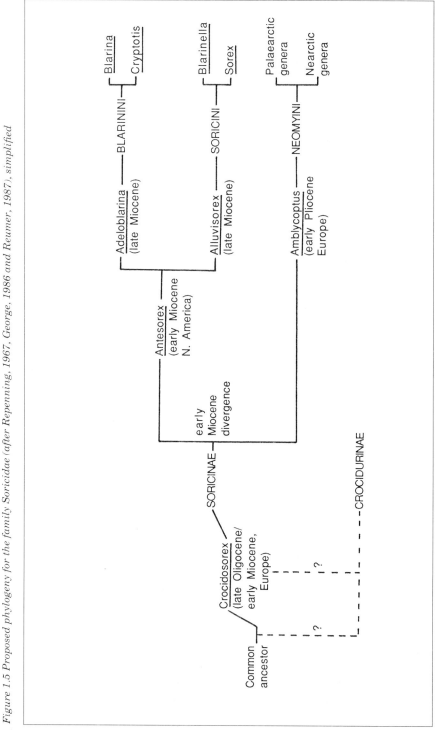

Figure 1.5 Proposed phylogeny for the family Soricidae (after Repenning, 1967, George, 1986 and Reumer, 1987), simplified

*Figure 1.6 Skeleton of the common shrew (*Sorex araneus*) showing the general body plan*

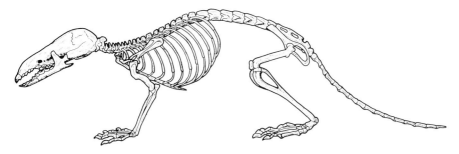

long toes (Vogel, 1974). The tails of shrews are generally covered with short bristles, and in *Crocidura* species these are interspersed with much longer hairs which may have a tactile sensory function (see Figure 1.7). As shrews such as *Sorex araneus* age, the tail becomes progressively less hairy and this provides quite a useful means of ageing individuals. Some species, however, have entirely naked tails. The base of the tail forms a useful storage centre in some mammals, and certain shrews exhibit this feature. Many species of *Crocidura* inhabiting arid regions of Africa, for example, have very fat tails which may help them through the dry season when food is in short supply.

*Figure 1.7 Tails of shrews: (a) water shrew (*Neomys fodiens*) with keel of stiff hairs; (b) common shrew (*Sorex araneus*), uniformly haired; and (c) white-toothed shrew (*C. russula*) with long bristle-like hairs*

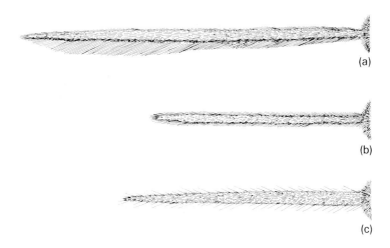

(a)

(b)

(c)

The feet of shrews show the greatest degree of adaptation to different modes of life, as indicated in Figure 1.8. This is particularly so with the water shrews, including *Neomys fodiens*, *Sorex palustris*, *Chimarrogale* and *Nectogale elegans*. They possess fringes of stiff, silvery hairs about 1mm in length on the sides of the toes and on the lateral edges of the feet on both the fore and hind limbs which increases the surface area of the

Figure 1.8 The feet of shrews showing adaptations for different modes of life (after Hutterer, 1985)

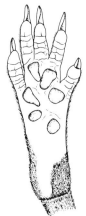

Ground-dwelling shrew
Myosorex eisentrauti

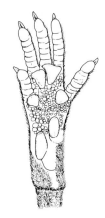

Scansorial (climbing) shrew
Sylvisorex megalura

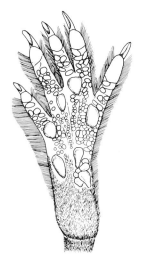

Semi-aquatic shrew
Neomys fodiens

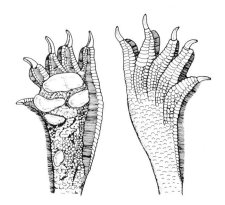

Semi-aquatic shrew
Nectogale elegans

feet and adds to the propellant power. These hairs fan out as the shrews spread their toes and paddle in the water. The American water shrew, *Sorex bendirii*, has been observed to run on the surface of the water for 3–5 seconds, apparently deriving its buoyancy from the air trapped amongst the hairs of the feet augmented by the surface tension of the water (Pattie, 1973). The feet of water shrews are also rather larger and broader than those of their terrestrial counterparts. These stiff hairs have another important function, namely in grooming the pelage following swimming. They act effectively as combs as the animals scratch and flick the water from the fur after a bout of diving.

14

Such fringes of stiff hairs provide an adaptation not only for swimming in water, but also for life amongst shifting sands, for the Turkestan desert shrew, *Diplomesodon pulchellum*, also possesses this modification, a remarkable example of convergent evolution.

Nectogale elegans, an Asiatic water shrew, possesses fringes of hairs on the feet which are rather more flattened than in other water shrews and, in addition, the feet are webbed (see Figure 1.8). Both fore and hind feet are fully webbed to the base of each terminal phalange. The palms of the feet are different from those of other shrews, too. Unlike the usual relatively small pads, those of *Nectogale* are large and disc-like and thought to act as adhesive surfaces to assist the grip on wet, slippery stones and rocks. They may even function underwater. The feet are covered dorsally by a series of small scales which become transverse scutes at their extremities.

The tails of water shrews also show some modification for swimming. In *Neomys* and *Sorex palustris*, the tail possesses a ventral keel of short, stiff hairs along its length. In *Nectogale* this adaptation is taken to the extreme, for there is not just one ventral keel but two, together with two lateral keels. This may assist the shrews to keep their balance as they swim along in the water, and prevent pitching and yawing.

Some of the anatomical modifications of shrews cannot be adequately explained, usually because not enough is known of their habits. For example, the armoured or hero shrew (*Scutisorex somereni*), a forest species found in northern Zaire, Rwanda and Uganda, possesses a vertebral column which is quite unique. The lumbar vertebrae number eleven rather than the customary five or six, are particularly strong and sturdy and bear a complicated system of articular apophyses or spines. These vertebrae are furnished not only with lateral interlocking spines but also with dorsal and ventral interlocking spines, creating a basket-like structure quite unlike any other mammal (see Figure 1.9). While it confers considerable strength, this is not a rigid structure and the vertebral column retains considerable freedom of both lateral and dorso-ventral movement. Such is the strength of this shrew that it is reported to be capable of supporting the weight of a full grown man weighing some 70kg balancing on one foot on its back. A shrew treated in this way was observed to run off unharmed as soon as the man stepped off its back (Allen, 1917). This extraordinary structure may be an adaptation for digging and burrowing amongst soil and leaf litter for food, but no other burrowing shrews have such an adaptation.

The Pelage

Shrews typically have a short, dense pelage which consists of three types of hair: outer guard hairs, awn hairs and soft, inner woolly hairs (Hutterer, 1985). But here again, there are quite distinct differences in the structure of the pelage according to the mode of life and the climate. Species inhabiting cool, temperate regions have longer and thicker coats than those living in hot, arid regions. Those such as *Crocidura lunosa* which inhabit montane forests have particularly long, soft coats. The dorsal hairs of some species of *Crocidura* are flattened and semi-spinous but the precise function of this adaptation is unknown.

*Figure 1.9 Skeleton of the armoured shrew (*Scutisorex somereni*) from Africa (a), with details of the thickened vertebral column (b), and close-up of vertebrae (c)*

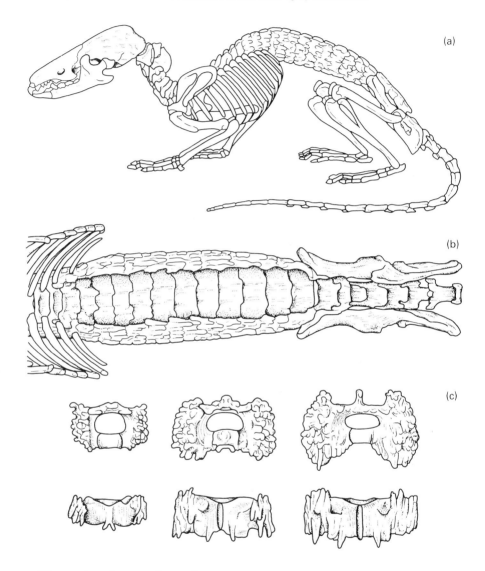

The water shrews show the greatest modification of their pelage for their particular lifestyle, which involves frequent submergences in water. Several of these species, such as *Chimarrogale*, *Nectogale elegans* and *Sorex palustris*, habitually plunge into ice-cold mountain torrents in which they forage for freshwater invertebrates, and their coats are not only adapted to resist wetting but also to offer increased insulation. The outer or guard hairs of shrews are not uniformly rounded in cross section but have indented sides forming lateral grooves which look roughly H-shaped in profile under the microscope (as shown in Figure 1.10). The lateral grooves contain ridges which increase in number and

16

Figure 1.10 Cross section of the hairs of shrews showing modifications for an increasingly aquatic lifestyle (after Hutterer and Hürter, 1981)

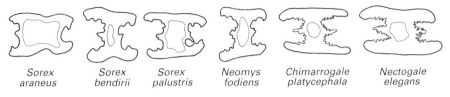

| Sorex araneus | Sorex bendirii | Sorex palustris | Neomys fodiens | Chimarrogale platycephala | Nectogale elegans |

Terrestrial - - - - - - - - - - - - - → increasingly aquatic

complexity from the terrestrial species such as *Sorex araneus* to the semi-aquatic *Neomys fodiens*, reaching their greatest complexity in *Chimarrogale* and *Nectogale* which are the best adapted for swimming and diving. The number of ridges in the lateral grooves of *Nectogale* is four times that of *Sorex araneus*. This peculiar structure is thought to act rather like a plastron, keeping air in the fur and so repelling water and providing increased insulation against heat loss. The ends of the awn hairs are also flattened in water shrews, providing additional water-repellent properties.

While most shrew species have a brown pelage, as a group they are far from uniform in colouration. The colour ranges from predominantly yellowish, reddish and greyish in species of *Crocidura* to dark brown in many temperate species such as *Sorex araneus*, *S. vagrans* and *S. monticolus*, and black in *Neomys fodiens*, *S. palustris*, *S. alpinus* and others. Some species even have a bluish or greenish hue. *Diplomesodon pulchellum* is almost piebald, with a grey dorsal surface with a large white patch in the middle of the back and white underparts, feet and tail. *N. fodiens* is particularly striking in colouration: most individuals have a black back, a white belly, a tuft of white hairs on the ears and a rim of white hairs encircling the eyes. Albino shrews are occasionally found, but more frequent is minor albinism on different parts of the body. For example, some 20 per cent of the population of *S. araneus* exhibit albinism of the ear tufts (Crowcroft, 1957), and there is albinism of the tail tip of this species, which has been found to vary geographically in Britain (Corbet, 1963).

The colouration of the pelage often provides some indication of the age of the shrew. For example, in *Sorex araneus* the juveniles are light brown until their first moult at around three months of age when this pelage is replaced by a rich, dark brown winter coat.

The Senses

While vision in shrews is poor, the olfactory, tactile and acoustic senses are generally well developed. Nevertheless, there are differences in the degree of development of these senses according to the mode of life, and the type of environment inhabited. The olfactory lobes of most shrews are large, but in water shrews such as *Neomys* which regularly hunt underwater, they are somewhat reduced. For these shrews, olfaction is of little help in prey detection and tactile senses are of more use. It is the muzzle with its covering of touch-sensitive vibrissae or whiskers which may be of particular importance in detecting vibrations or small movements of prey. The vibrissae are particularly well innervated in aquatic shrews by the enlargement of the trigeminal nerves in the muzzle, and this in turn has resulted in an overall increase in the size of the brain and the cranium or bony case surrounding it compared with more terrestrial shrews such as *Sorex araneus* (see Figure 1.11). Those shrews most adapted to an aquatic life, namely *Chimarrogale* and *Nectogale* show this feature most strongly. Just how important the vibrissae are in prey detection will be discussed further in Chapter 5.

Figure 1.11 Variation in the skulls of shrews having different modes of life. Half the skull is viewed from above and behind. Note the progressive broadening of the brain case and the enlargement of the foramen magnum from the terrestrial Sorex to the semi-aquatic Nectogale (after Hutterer, 1985)

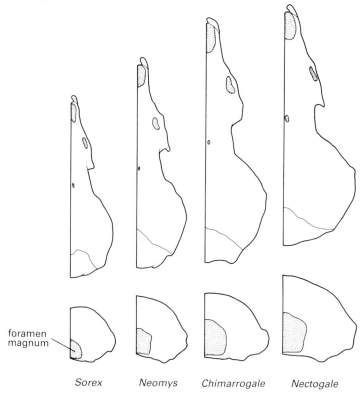

In addition to the copious covering of vibrissae on the snout, other long whiskers are present on the head, as in most other mammals. Mystacial vibrissae on either side of the snout and submental vibrissae below the jaw may help the shrew to judge the sizes of holes and crevices through which it squeezes or the nearness of objects in its path. Each forelimb also has a single antebrachial vibrissa behind the wrist area which may help to provide a sense of distance between the limb and the substratum.

The snout of a shrew is characteristically long and pointed and it is in

Figure 1.12 Variation in the form of the rhinarium (nose) in shrews having different lifestyles (after Hutterer, 1985)

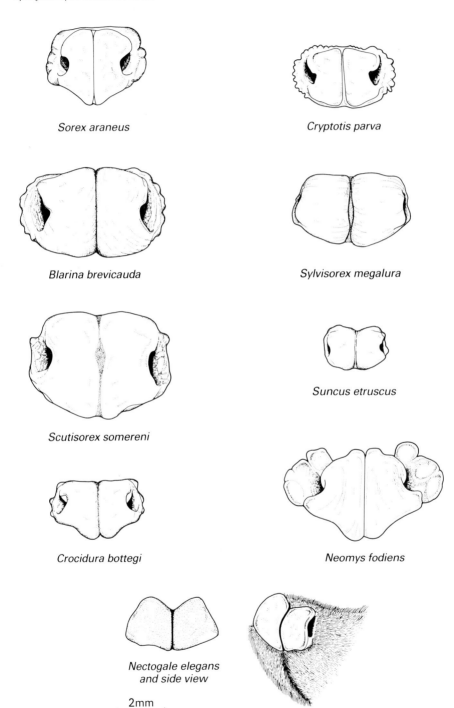

Sorex araneus

Cryptotis parva

Blarina brevicauda

Sylvisorex megalura

Scutisorex somereni

Suncus etruscus

Crocidura bottegi

Neomys fodiens

Nectogale elegans
and side view

2mm

constant motion, wiggling rapidly to and fro, sniffing the air and probing the substratum. At its tip is a moist, highly sensitive, smooth glandular pad or rhinarium. Again, the form of the rhinarium shows some variation in shrews with different lifestyles. Figure 1.12 shows examples of the rhinaria of shrews: in *Neomys fodiens*, which is semi-aquatic, the lateral lobes of the rhinarium are very large and are thought to act to close the nostrils while the shrew is swimming underwater. In the semi-aquatic *Nectogale* and *Chimarrogale*, the nostrils are situated behind the rhinarium pad, again to prevent the entry of water.

Olfaction not only plays an important part in prey detection but also in the social life of shrews. They have a number of scent glands associated with the skin on various parts of the body, including the neck, throat, below the base of the tail, beside the lips, behind the ears and even on the soles of the feet, but the most prominent are those situated on the flanks,

Figure 1.13 Variation in the form of the ears of shrews adapted to different modes of life (after Hutterer, 1985)

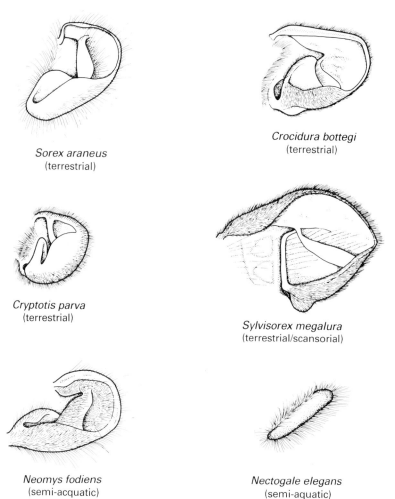

Sorex araneus
(terrestrial)

Crocidura bottegi
(terrestrial)

Cryptotis parva
(terrestrial)

Sylvisorex megalura
(terrestrial/scansorial)

Neomys fodiens
(semi-acquatic)

Nectogale elegans
(semi-aquatic)

20

midway between the fore and hind legs. These lateral flank glands are small, oval areas bordered by stiff hairs. They are well vascularised and contain many sebaceous and sweat glands which, when the shrews are mature, produce a greasy, highly odoriferous secretion which rubs off on objects as the animal brushes past. More will be said about their role in the life of shrews in Chapter 4.

The acoustic sense of shrews is also important. They are very vocal animals and they both perceive and emit a wide range of sounds including ultrasounds, about which more will be said in subsequent chapters. While shrews typically have small external ears, their size and form are quite variable, as Figure 1.13 demonstrates. Again, this can be related to habitat and mode of life. Scansorial species and those inhabiting the ground surface possess the largest and most prominent external ears or pinnae; those from temperate regions have quite hairy ears which are mostly hidden amongst the fur, while those species originating in tropical regions have large and nearly naked pinnae. Species which are semi-fossorial or semi-aquatic, on the other hand, tend to have much smaller ears situated more to the rear of the head and often obscured by the pelage. This is taken to the extreme in *Anourosorex* (semi-fossorial) and *Nectogale* (semi-aquatic) in which the ear pinnae are entirely lacking and there is merely a narrow opening to each ear and a slight folding of the skin on the side of the head (see Figure 1.13).

In contrast, sight seems to be of very limited importance to shrews. Unlike the large olfactory lobes found in their brains, the optic regions are small and poorly developed. The eyes themselves are extremely small, the eyeball being only 0.7–1.5mm in diameter, although the structure is analogous to that of mammals possessing good visual acuity (Branis, 1981). The prime function of vision in shrews seems to be concerned simply with recording light intensity (Branis, 1981), although vision probably varies from species to species, for some possess an efficient retina (Grün and Schwammberger, 1980) while in others the eye appears to be non-functional. In *Nectogale*, for example, there seems to be no opening to the eye at all, and it is covered by skin (Hutterer, 1985).

A summary of the major anatomical features of shrews adapted to different modes of life is given in Table 1.3.

CHROMOSOMES

Shrews exhibit certain peculiarities with respect to their chromosomes. In most mammals the number of chromosomes is constant and is characteristic for each species. Shrews, however, show considerable variation in chromosome number, through a phenomenon known as Robertsonian polymorphism, although the full extent of this is not known. Ford *et al.* (1957) found the number of chromosomes of *Sorex araneus* to vary from 21 to 27 in males and from 20 to 25 in females. They also described a multiple sex chromosome mechanism with males being XY_1Y_2 and females XX. The biological significance of this phenomenon is not quite clear but it does make possible a large number of different chromosome patterns, many of which have been found to occur in wild populations.

Table 1.3 Summary of the major anatomical adaptations of shrews adopting different modes of life (for further explanation see text)

	Terrestrial	Scansorial	Psammophilic	Semi-fossorial	Semi-aquatic
PELAGE	Short & dense; hairs may be flattened & semi-spinous	—	—	Very short & dense	Denser than in non-aquatic forms; awn hairs with pronounced lateral grooves & flattened tips
FEET	Feet & claws simple; short to medium in length	Elongated feet & toes	Elongated claws; feet fringed by stiff hairs	Elongated claws on fore feet	Feet & toes fringed by stiff hairs; feet may be webbed & possess disc-like pads
TAILS	Medium to long, covered with bristles; may have fat deposit at base	Elongated	Shortened	Shortened	Long, with keels of stiff hairs
EARS	Medium to large	Large	—	Reduced in size; situated towards rear of head; often obscured by pelage; pinnae may be entirely lacking	
EYES	– – – Small but visible amongst fur – – –			– – – Extremely small & largely obscured by fur	Eyes may be covered by skin
NOSES	Form of rhinarium very variable	—	– – – No special adaptations evident – – –		Rhinarium with pronounced folds of skin & nostrils may be closeable
BRAINS	– – – Olfactory lobes well developed; optic regions less so – – –				Olfactory lobes reduced; enlargement of cranium & trigeminal nerve

Shrews also form distinct chromosome races. In Britain, three races of the common shrew (*Sorex araneus*) have been described (Searle, 1986; Searle and Wilkinson, 1987). There is an 'Aberdeen' race on the northern and western periphery of Britain, an 'Oxford' race in the central and eastern region, and a 'Hermitage' race which is more intermediate in range, supporting the hypothesis that races of shrews spread into Britain in successive waves at the end of the last glacial period.

OTHER ATTRIBUTES OF SHREWS

Shrews are characterised by voracious appetites and high metabolic rates. The respiration rate of a shrew can exceed 5,000cm^3 of air per kilogram body weight per hour, compared with 3,500cm^3 for a mouse of similar size and 200cm^3 for man. The activity and energetics of shrews will be discussed more fully in Chapter 6. Nevertheless, they are robust animals which are active night and day, winter and summer. They are resistant to cold and heat but not to starvation: many species, particularly those in northern temperate regions, must feed every two or three hours or they will die. Contrary to popular belief, they are not highly strung nervous animals which die easily from shock, although at times of alarm, excitement or great activity, their heart rate is reported to increase from 88 to some 1,320 beats per minute.

2 The life history of shrews

BREEDING

Sexual Maturation and the Onset of Breeding

Most shrews in northern temperate regions pass the winter in an immature state, and the reproductive organs of both males and females are small and non-functional. Occasionally, young females breed in their first calendar year when they are less than five months old, but this is rare in soricines. Both sexes are about equal in size through February and the beginning of March, although males may be slightly larger. Then, as spring comes, both sexes undergo a very rapid change to sexual maturity, with males reaching maturity earlier than females, often by some three weeks. The increase in body weight at this time is accompanied by an increase in body dimensions including the skeleton and, most obviously, the reproductive organs. In males the change is particularly marked: the paired testes which in overwintering immatures each measure a mere 1.5–2.0mm in length and are well hidden in the abdominal cavity, grow enormously to 7.0–8.0mm in mature common shrews (*S. araneus*) and are obvious externally as large inguinal swellings on either side of the anus (see Figure 2.1). The penis, too, becomes greatly enlarged and thickened. Most males are fully mature by April and sometimes earlier: I have caught a mature male in late February in southern England. Males at this time are heavier than females.

Females grow especially fast in April when they mature. The reproductive tract is T-shaped with a single vagina branching into a pair of uterine horns at the top of the T, each horn leading to a small ovary (see Figure 2.1). In immature females the reproductive tract is small and thin with the combined length of the uterine horns in *S. araneus* being a mere 4–7mm. But the vagina and uterus enlarge and thicken greatly at maturity, such that the length of the uterine horns reaches 17mm in breeding condition. The nipples also become enlarged and more clearly visible at this time.

Accompanying the growth to maturity is a change in behaviour. The shrews become much more active on the ground surface and they begin to squeak vociferously at each other when they meet. While their presence

Figure 2.1 Reproductive organs of male and female shrews in breeding condition

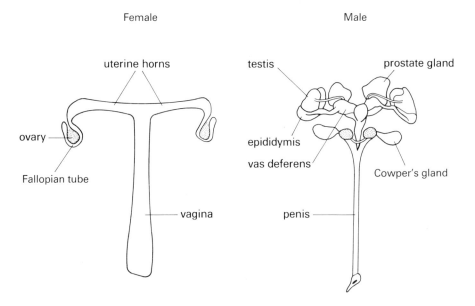

is difficult to detect over the winter period, suddenly in spring they are easily heard by their high-pitched squeaks and chatters in hedgerows and fields as males abandon their winter home ranges and wander widely in search of females, crossing the home ranges of neighbouring males and females as they do so. They also become much more aggressive at this time, particularly the males. This is noticeable when mature wild shrews are handled: in spring and summer they kick, bite, thrash their tails and utter angry churls while at other times of year they are more docile.

The rapid changes undergone by shrews in northern regions as they mature are the result of increased hormonal activity, but the cues for these changes have not been fully determined and experimental evidence is conflicting. It does seem that the increasing length of the photoperiod as the weeks pass from winter to spring is the most likely cue, as it is in birds. Changes in ambient temperature have a minimal effect, for shrews come into breeding condition even if there is snow on the ground in March and April. However, Skaren (1973) did notice that the earliest pregnant *Sorex araneus* he caught in late April coincided with a rise in ambient temperature and the loss of most of the winter snow in Finland where his study took place.

By late April and May northern shrews are breeding, and the first litters are born. Ninety per cent or more of females will be pregnant and/ or lactating by the end of May. Females have a post-partum oestrus and approximately one-third of the mature females of *S. araneus* will become pregnant again immediately after parturition (the birth of their first litter), so they will be simultaneously lactating and pregnant (Skaren, 1973). Juveniles first appear in the population, having vacated the parental nest, in early May.

While the morphological and physiological changes which accompany sexual maturation are similar in all shrews, the onset of breeding may occur at different times of year according to the geographical range of each species. The breeding season of the African mouse shrew (*Myosorex varius*) coincides with the rainy season in the Drakensburg of South Africa. It matures between July and September and then breeds right through to the following March. Breeding reaches a peak from September to November, coincident with the onset of the spring rains, and then breeding activity declines. Temperature also seems to be a factor affecting the timing of breeding in this shrew, for those at higher altitudes (above 2,200m) commence breeding later than those at lower altitudes (below 1,900m). Although the rainfall is greater at higher altitudes in the Drakensburg, the temperature is lower and may cause breeding to be delayed.

Crocidurine shrews from southern and tropical regions may reach sexual maturity at a much earlier age than most soricines. They can mature within a couple of months or so of weaning, and commence breeding in the calendar year of their birth. They continue to breed throughout life, although for most species breeding will be interrupted by the dry season.

The Breeding Season

Shrews in temperate regions have a well-defined breeding season in the wild. Generally speaking, they attain sexual maturity in the spring of their second calendar year, and breeding occurs in late spring and summer, finishing by autumn at the latest. *Sorex araneus*, *S. minutus* and *Neomys fodiens* in Britain and Europe, for example, have a breeding season extending from April to September, although most young are born in June, July and early August. In Canada and the northern USA, most species breed from spring to late summer. Under more favourable climatic conditions, the breeding season may be extended. For example, most pregnant females of *Sorex vagrans*, *S. ornatus* and *S. sinuosus* were found in March–April by Rudd (1955) in California, USA, but he also recorded some as early as February and as late as the last week in October. By commencing to breed early in the season, more litters may be produced. Strangely, even with such a prolonged breeding season he found little evidence that more than one litter per female was produced. He also concluded that young born late in the summer may themselves not breed until late the following summer.

Godfrey (1978) carried out a detailed study of breeding in captive colonies of *Crocidura suaveolens* on the Channel Island of Jersey where shrews were kept under conditions very close to ambient temperatures but were fed *ad libitum*. She found that the first litter was born in mid-April and the last in late October, but breeding reached a peak in May when most litters were produced and some 83 per cent of mature females were pregnant or lactating. Since first conception in March, breeding in one colony continued for eight months. In another colony, the first litter was produced in late December and the last in the following September, so that breeding continued for ten months, and was able to take place even in winter. She concluded that the main breeding season extended for 8–9 months from February to October but that it could be extended under certain conditions (such as increased food supply). This apparently mirrors wild populations of this species: in the Scilly Isles, *C. suaveolens* has a similar breeding season and winter breeding has been recorded in some years (Rood, 1965), although this phenomenon appears to be unrelated to climatic factors. The breeding season of *C. russula* both in the Channel Islands and in France is also February–October. Winter breeding in wild *Sorex* in northern temperate regions has rarely if ever been recorded.

In more southerly climes and in tropical regions the breeding season of shrews may be extended. *Notiosorex crawfordi*, the desert shrew, breeds from April through to November in the south-western USA. *C. russula* in the Mediterranean region begins to breed slightly earlier than its more northerly counterparts, and may prolong breeding in the autumn. Breeding can occur throughout the year in many wild crocidurines, including *C. fuliginosa* in south-east Asia (Medway, 1978) and several *Crocidura* species in Zambia (Sheppe, 1973). However, even in these species breeding reaches a peak at certain times of year, usually coinciding with environmental conditions. For instance, *Suncus murinus* can breed all year round, but in the Malay peninsula most pregnancies have been recorded in October–December (Medway, 1978); in India this species breeds mostly during the monsoon months when food is plentiful, and in Nepal it breeds in April–September. *Cryptotis* is rather an unusual soricine. It probably breeds throughout the year in the southern part of its range but is restricted to a March–November season in northern areas. *C. parva* shows no evidence of a true oestrous cycle: mature females are sexually receptive 24–48 hours after being placed with males and they will mate again only 1–4 days after giving birth (Mock and Conaway, 1975). This may also be a feature of crocidurines, which can breed continuously, but apparently not the other soricines with their more restricted breeding.

Courtship and Mating

Shrews do not exhibit a well-defined sequence of behaviours which culminates in mating as do many other mammals. Courtship is rudimentary: once a mature male has located a female in breeding condition he pursues her relentlessly until contact is made. Chasing occurs, with the male following the female very closely, often with his nose over her rump. Now and again the female stops and permits the male to sniff and mount her. In the African musk shrew, *Crocidura flavescens*, behaviour known as flehmen has been observed, in which the male curls back the top lip and sniffs at the female (Baxter and Meester, 1982). This lip-curling behaviour is more commonly associated with large herbivores such as antelope and horses as they test the odours produced by females.

The male's advances are often met with fierce rebuffs on the part of the female, for unless she is in oestrus she will not mate. Non-oestrous females will drive away courting males by uttering loud squeaks, and they may attempt to attack and bite persistent males. In the common shrew (*Sorex araneus*), oestrus lasts for only a few hours every three weeks or so in the reproductive cycle of the female, and this is the only time at which she will permit copulation to occur. Even when the female is receptive, mating is accompanied by scuffles and vocal churls and squeaks. Mating is brief: the male mounts the female and uses his incisor teeth to grip her by the nape of the neck, or more frequently, by the fur on the top of the head. This often results in tufts of hair being removed, and mated females can easily be identified by the small naked patch on the top of the head. Copulation lasts from a few seconds to around half a minute in *Sorex araneus*, and may occur repeatedly until ejaculation has occurred. After this the female aggressively drives away the male and he shows no further interest in her but wanders off in search of other mates.

In crocidurine shrews the male may stay with the female and her young and help in gathering nesting material, in nest-building and even in tending the young, as has been reported in *Crocidura russula*, *Suncus etruscus* and *S. murinus*. Females of most species do not usually tolerate the presence of males in the nest with the young, but this is not so with *C. russula*. Here, the male is permitted to stay at the nest and will crouch over the young, even when the female is out foraging. Males of *C. russula* have even been found to retrieve scattered youngsters, but this seems a rare occurrence. The extra body heat provided by males as they remain in the nest may have a very useful role in the survival of the naked youngsters, and permit the female to spend longer periods out foraging and replenishing her resources. The extent of male parental care in other species of shrew is not known.

Some species of shrew have been found to emit quite distinct vocalisations during courtship. For example, the male water shrew (*Neomys fodiens*) utters a series of pure tones while following the female, and she produces chirps which may indicate her receptivity when meeting a male (Hutterer, 1978). Similar sounds have been reported in a number of other species, including *Crocidura russula*, *C. suaveolens*, and *Suncus etruscus*. Clicking sounds accompany the courtship of *Blarina brevicauda* (Gould, 1969). These vocalisations may serve to reduce the normal aggressive responses of shrews when they meet each other, or provide an indication of the intentions of the male and the state of receptivity of the female.

Another feature which may also affect the receptivity of the female is the state of the lateral flank glands of the male. While these scent glands are present in both males and females of most species of shrew, they seem to be only functional in the males, and only during the breeding season. The development of the flank glands and the production of odoriferous secretions from them occurs simultaneously with the rapid enlargement of the testes and other reproductive organs of the males, and so they may have a role in communicating their breeding condition to females, or even to other males. However, little is yet known about the precise function of these glands. Scent communication is discussed further in Chapter 4.

The act of copulation seems to stimulate ovulation in the females of those species whose reproductive biology has been studied in most detail, including *Sorex araneus*. Pearson (1944) demonstrated that repeated copulation was necessary to induce ovulation in the short-tailed shrew, *Blarina brevicauda*. Fertility and the rate of pregnancy appear to be very high amongst mature shrews, for 80 per cent or more of females are either pregnant or lactating soon after sexual maturity is attained. Most females seem to become pregnant, then, at their first oestrus.

Gestation Period

The gestation period of shrews is similar to that of small rodents of like size, but it varies slightly from species to species. The most precise assessment of gestation in shrews was made by Godfrey (1979) with a captive colony of common shrews (*Sorex araneus*). Each day she took vaginal smears from the mature females to assess their sexual condition, and from these she could tell when mating had occurred. She concluded that gestation lasted 24–5 days in *S. araneus*. Occasionally, however, pregnancy may occur while these shrews are still lactating, as a result of mating and conception during the post-partum oestrus. In this case, gestation may be prolonged to some 27 days (Vogel, 1972) to prevent both litters of young being nursed at the same time. It has been suggested that delayed implantation may occur in some species of shrew, but there are no definitive accounts of it.

Perhaps rather surprisingly, the smaller pygmy shrew (*S. minutus*) has a marginally longer gestation period than the common shrew, amounting to at least 25 days (Hutterer, 1976) while the larger water shrew (*Neomys fodiens*) and the short-tailed shrew (*Blarina brevicauda*) both have slightly shorter gestation periods of about 20 days. It seems that the larger the size of the shrew, the shorter is the gestation period. However, the Crocidurinae tend to have slightly longer gestation periods than the Soricinae: Vogel (1972) recorded pregnancies lasting 28–33 days with a mean of 30 days in *Crocidura russula*, and it is also approximately 30 days in *Suncus murinus*. This is to some extent correlated with the rather more advanced state of development at which the young of crocidurine shrews are born.

Birth of the Young

As pregnancy advances and the birth of the young approaches, the female's abdomen distends markedly and her body becomes pear-shaped. A few days before the birth of the young, she builds a dome-shaped nest which is much larger and more complex in structure than the usual nest used for resting and retreat day by day. The nest is composed of a mixture of dried grasses, leaves, moss and any other suitable materials that may be available, and it is placed beneath a fallen log, in a grass tussock, or in an underground burrow. Captive individuals will readily adopt small boxes as nesting sites. The female will collect mouthfuls of nesting material, carry them to the nest site and arrange them in a large pile. She sits in the centre of the nest-to-be and arranges the material carefully around her, handling individual fragments with her mouth and probing with her snout, gradually weaving them together to form a discrete structure which has two or more entrances. The feet are rarely used in nest-building, but a rounded, open interior to the structure is created by frequent shifts of the body pressing against the inside walls. The nest may measure 12–15cm in diameter. While in most species it is only the female that builds the nest for the young, in *Suncus murinus* (the large Asian musk shrew) both sexes collect material for the nest.

Within the nest the young shrews are born naked and blind. The young of common shrews (*Sorex araneus*) each weighs approximately 0.5g at birth, with a head and body length of 15mm; those of the smaller pygmy shrew (*S. minutus*) weigh a mere 0.25g. Crocidurine shrews, such as *C. russula*, are slightly larger and better developed at birth than soricine shrews, and in *C. russula*, which is similar in adult size to *S. araneus*, the newborn youngsters weigh up to 0.9g. The mother assists the young during the birth, biting through the embryonic membranes and umbilical cords and eating them along with the placenta.

Litter Size

The number of young per litter varies considerably, not only from one species to another but also between individuals. The maximum number is 9–11 young per litter in the majority of species, although the water shrew (*Neomys fodiens*) has been known to produce 15 young in a single litter. The usual litter size for many species, including *Sorex araneus*, *S. fumeus*, *S. palustris* and *N. fodiens*, is 5–7 young, but crocidurine shrews such as *C. russula*, *C. suaveolens* and even the large *Suncus murinus* tend to average 3–4 per litter. Generally, the average litter size for the Soricinae is more than five while that for the Crocidurinae is fewer than five, reflecting the small differences in gestation periods and development of the young at birth between these two groups of shrews.

Not all the young in a litter will survive, although the young born into the smaller litters, such as those of crocidurine shrews, have the best chance of success. Not all shrews have the same number of nipples, and this feature may be reflected in the number of offspring born per litter which survive. The nipples are inguinal, situated on either side of the abdomen. While most species have three pairs of nipples (including species of *Sorex*, *Blarina* and *Crocidura*), some such as *Sorex hoyi* have four pairs of mammae, and *Neomys* has five pairs. As mentioned above, *Neomys fodiens* is capable of producing up to 15 young in a single litter, but how may survive in such large litters is not known.

Competition between shrews for resources probably begins in the womb, since there are small differences in size between individuals at birth, and this continues after they are born. In the larger litters there are always runts which fail to survive long. During the period of suckling, the differences between litter-mates in terms of their body weights, general condition and the rate of development increases as the larger ones compete successfully for the limited number of teats at which to feed, and the smaller ones fail and die. Even in those species which have five pairs of nipples, such as *Neomys fodiens*, a female cannot realise her full potential by rearing all the young she produces in a single litter. Usually only three or four survive to weaning.

Development of the Young

Despite their poorly developed and helpless state at birth, young shrews grow very rapidly, particularly in the first 10–15 days of life, as can be seen in Figure 2.2. During their first few days, the movements of the young in the nest are confined to suckling and rolling to and fro, although body movements are already quite well co-ordinated. They utter a mixture of cries and whistles when distressed, particularly in the absence

Figure 2.2 Growth and development of young common shrews (Sorex araneus)

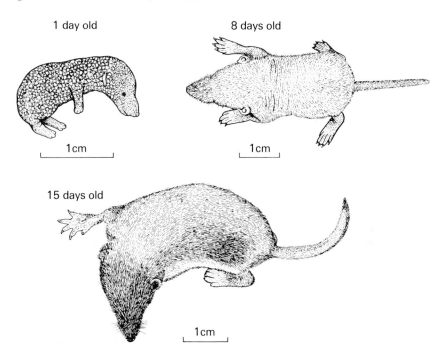

of the female. At about nine days, the fur of young common shrews (*S. araneus*) appears as a soft grey down, and by eleven days, the teeth show their characteristic yellow tips.

By 14 days they have achieved body weights of 5–7g and are covered in a short, fine, greyish fur, and the eyes are just beginning to open, but they are still confined to the nest. They emit loud, relatively low-frequency cries resembling barks when distressed or hungry, and they will exhibit escape behaviour for the first time if they are disturbed, crawling deeper into the nest or even vacating it altogether. As with the offspring of other mammals, these young shrews show typical contact behaviour, crawling over and underneath each other, especially if placed in strange surroundings. This behaviour may persist for several weeks if the young are kept together.

By 16 days their eyes have fully opened, and by 18 days the youngsters are nearly as large as their mother, and their body weights cease their exponential increase and level off (see Figure 2.3). At this stage they may venture out of the nest for short periods independently of the mother, who may drag them back to the nest if she discovers them missing. The young are still suckling at this time, and in order to feed them, the female increases her own food intake to some 125 per cent of her own body weight per day (compared with about 80 per cent prior to pregnancy). I first observed captive common shrews to take solid, invertebrate food at 21 days old, but they still made attempts to suckle, although this was discouraged by the mother who responded with threat calls and angry churls.

Figure 2.3 The pattern of growth of young common shrews (Sorex araneus) *from birth to weaning*

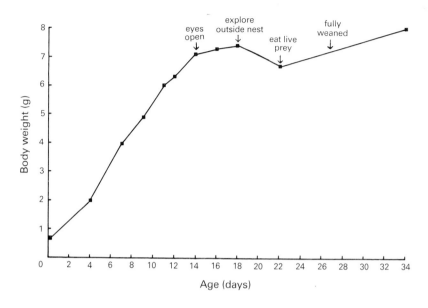

During the weaning period, which lasts about five days, the young shrews lose 1–2g in weight, but as soon as they are fully weaned at 22–25 days old, they start to gain weight once more. They then begin to disperse and occupy separate nesting sites wherever suitable cover and bedding material is available. However, these inexperienced shrews construct only rudimentary nests for themselves at this stage. If they are confined together in an enclosure, aggressive behaviour between the young and between them and the mother is very apparent: squeaks and scuffles occur both in the nests and in the open enclosure wherever they meet, and competition for nests and for food occurs. At approximately 25 days after birth young common shrews are completely independent.

The rate of growth and development of the young appears to be affected by environmental conditions, particularly temperature, for development is retarded at lower temperatures. Weather conditions could, therefore, be a critical factor in the success rate of the females in rearing their young. Another important factor is the size of the litter, as suggested above, since females often produce many more young than can be reared.

When the young are first born, the female remains with them, venturing out only for brief foraging trips. She reacts very aggressively to intruders, biting and uttering loud warning churls. Captive shrews often move their young to new nesting sites, and even those housed in outdoor enclosures provided with soil, leaf litter, grass and logs will exhibit this behaviour. The nests do become soiled and damp with use, which may provide the stimulus to move to a new site, and disturbance will also encourage relocation of the nest. When very small and helpless, the young are carried to the new nest by the mother in her mouth. As they grow too large to be carried, they are dragged by the scruff of the neck or pushed along by the mother. A particular behavioural trait associated with shrews at this stage of their development is caravanning. This behaviour has been described in captive crocidurine shrews such as *Crocidura suaveolens* and *C. russula* (Corbet and Southern, 1977), *C. bicolor* (Ansell, 1964), *Suncus etruscus* (Fons, 1974) and *S. murinus*. Each youngster grasps the base of the tail of the preceding shrew so that the mother runs along with the young trailing in a line behind her (see Figure 2.4). It seems to be stimulated by disturbance of the nest and occurs as the shrews move to a new site, but it may also be used to encourage exploration and extend the youngsters' knowledge of the environment outside the nest as weaning approaches. In *S. murinus*, the large Asian musk shrew, caravanning behaviour in captivity occurs between ten and 22 days after birth of the young, with a peak around 16 days. Harper (1977) graphically describes caravanning in a party of seven wild soricine shrews, probably *Sorex araneus*, which she observed for some 15 minutes before they disappeared from sight into the undergrowth. They ran repeatedly in a broad figure-of-eight around the bases of two hawthorn bushes, in the manner described above, with the lead shrew (presumed to be the mother) keeping up a continuous twittering throughout. This was the first report of caravanning both in wild shrews and in *Sorex*.

Figure 2.4 Caravanning behaviour in young crocidurine shrews

Young sometimes become displaced from the nest, either by rolling out when they are very small or by wandering out as they grow older. The mother is apparently not aware of the precise number of young in a litter and does not react if one youngster is missing until she hears it uttering distress calls, whereupon she immediately commences a search for it and carries or guides it back to the nest. Shrews seem to exhibit this behaviour quite indiscriminately, to the extent of retrieving and even fostering young which do not belong to them. Not only will they do this for members of their own species but also for young of entirely different species. A captive female common shrew (*Sorex araneus*), already nursing a litter of her own, responded immediately to the distress calls of six three-day-old, orphaned, white-toothed shrews (*Crocidura suaveolens*) when they were placed in her enclosure. She quickly located them and then carried them back to her nest and suckled them along with her own young. Fostering proceeded successfully for some three days, but then the mother died and the young could not be maintained.

Survival of Young

Young are most vulnerable at three stages during their post-natal development. The first week after birth is particularly critical, especially for those in a large litter of 6–9 young. As has already been pointed out, shrews frequently produce large litters containing more young than there are nipples available, so that competition for food occurs from a very early age. Within the first week after birth, the growth of the young is very rapid and variable between individuals such that the smallest and weakest succumb. The mother seems to show no interest in the small, sickly runts of the litter. Two or three young usually die at this stage in soricine shrews, unable to compete with larger, stronger siblings. Once the litter has been reduced in this way, the remaining young grow at a more even rate and have a good chance of survival until weaning.

Weaning represents the second vulnerable state, when young may decrease in weight for a few days in the transition to a diet of solid invertebrate food. At this stage the female can no longer sustain several large offspring on a diet of milk, and she discourages them from suckling. Exactly how weaning is achieved is not known but young shrews have been observed to lick at the mouth of the mother (a process known as lip-licking) which may stimulate her to regurgitate partially digested food for them, as happens in some other mammals. The female shrew has never been observed to catch food for her young so it seems that they instinctively go off and forage for themselves. It is possible that they accompany the mother in their first foraging expeditions, but observations of captive shrews suggest this is not so. The young have little fat reserve to sustain them during this critical period, and they will die if they go without food for more than about three hours.

The process of becoming independent and dispersing to new home ranges and nesting sites represents the third stage of vulnerability for young shrews. Not only may competition for food and nesting sites be an important factor in their survival (contributing to agonistic and even aggressive interactions between siblings, other young shrews and adults), but predation of these young, inexperienced shrews as they search for and establish suitable home ranges probably leads to a high mortality rate.

The Number of Litters Produced

Field studies of species such as *Sorex araneus* and *S. minutus* suggest that these shrews produce one or two litters per female per breeding season, and that most females have finished breeding by late July or early August. Occasionally a third litter may be born, or some females may breed late, so that young may be produced in late August and early September. In captivity, the maximum number of litters produced per female may be increased. For example, Vogel (1972) recorded a maximum of four litters per female in *S. araneus*, three in *Neomys fodiens*, and an incredible eight litters during the life time of *Crocidura russula*. These crocidurine shrews tend to live longer than soricine shrews and, once they have reached maturity, they can be stimulated to breed almost continuously when kept in small groups in captivity under optimum conditions of temperature, light and food supply.

The period of fertility may last just over one year in these crocidurine shrews (Vogel, 1972) which is significantly longer than in most soricine shrews such as *S. araneus* and *N. fodiens*, at least in captivity. They are also capable of reaching sexual maturity much earlier than the majority of soricines. Vogel (1972) found that 75 per cent of female *C. russula* reached sexual maturity within three months, the earliest being 58 days in females and 71 days in males. *Blarina brevicauda* can attain maturity at about 44 days in females and 84 days in males. But *Sorex araneus* reaches maturity only after 4–6 months, even given optimal conditions of temperature, light and food supply, although increased daylength in the late winter and early spring can hasten sexual maturation (Crowcroft, 1964).

It can be seen, then, that the soricine and crocidurine shrews differ considerably in reproductive strategies, gestation period, litter size, condition of the young at birth, number of litters produced, length of the breeding season, onset of sexual maturity, and even the role of the male in helping with the nest and the rearing of the young. This probably reflects their different evolutionary histories, particularly with respect to the different environments to which they have become adapted: the crocidurines are essentially tropical and the soricines temperate in distribution.

LONGEVITY AND LIFE-EXPECTANCY

Most shrews are essentially annuals. Following birth in the summer, most temperate shrews overwinter as immatures, breed in the following spring and then die. This is reflected in the seasonal cycle in population numbers exemplified by the common shrew (*Sorex araneus*) in central England and shown in Figure 2.5. Here, there is a swell in numbers in summer as the juveniles are born, a marked decrease in autumn as old

*Figure 2.5 Seasonal population numbers of the common shrew (*Sorex araneus*) in scrub grassland in central England (after Churchfield, 1980a)*

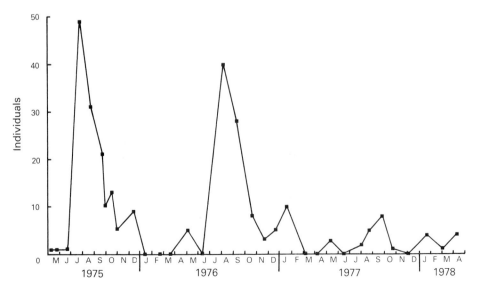

adults die off, and the surviving young shrews then carry the population through the winter. For most shrews, the life cycle is completed in little over one year, although the life-span does differ from species to species. The life-span of *S. araneus* in the wild is usually 12–13 months but occasionally individuals do survive longer: the maximum life-span recorded by Shillito (1963a) during three years of study was 16 months, and by me during eight years of study in different habitats 17 months. Most adults in the population have died by September in their second calendar year, but a few survive until November.

North American shrews of similar size have life-spans in the same order as *S. araneus*. For example, the smoky shrew (*Sorex fumeus*) lives for 14–17 months; *S. ornatus*, *S. vagrans* and *S. sinuosus* have reported maximum life-spans of 16–17 months (Rudd, 1955). The mouse shrew (*Myosorex varius*) from southern Africa resembles these shrews in size and habits, and is also an annual. Adults die off following breeding, approximately a year after their birth. Even very small shrews have life-spans of just over one year: in wild *S. minutus* it is some 13 months.

Larger shrews have a slightly extended life-span. For example, *Neomys fodiens* can live for 14–19 months in the wild (Dehnel, 1950; Price, 1953) and *Blarina brevicauda*, the short-tailed shrew, lives for about 2.5 years. *Suncus murinus* and the other giant, rat-sized shrews can probably survive even longer, but no long-term studies have been made of these species so little is known about their biology.

Body size is not the only factor affecting longevity, for the crocidurine shrews are notable for their greater life-spans compared with the Soricinae. The white-toothed shrew *Crocidura russula*, for example, lives about twice as long as *Sorex araneus* or *Neomys fodiens* (Vogel, 1972) and so may live through a second winter in the wild. The maximum recorded life-span for this species and *Crocidura suaveolens* in captivity is four years. *Cryptotis parva*, which is similar in size, lives for a maximum of 31 months in captivity (Mock and Conaway, 1975). The difference between these two groups of shrews is attributed to their metabolism: the Crocidurinae have lower metabolic rates than the Soricinae, and this may permit greater longevity.

The life-span of shrews can be prolonged to some extent by maintaining them under optimum conditions of temperature and food in captivity. *Sorex araneus* and *S. minutus*, for example, can live to 16 or 18 months or longer in captivity compared with the usual 12–13 months in the wild; *Crocidura russula* and *C. suaveolens* can live up to four years in captivity compared with 1.5 years in the wild, but the difference in maximum life-span is generally not very significant. However, it is the average life-expectancy which is greatly enhanced by captive conditions where there are no predators or competitors for resources. Figure 2.6 shows a survivorship curve for *Sorex araneus* and *Neomys fodiens* compiled from the results of a live-trapping programme and mark-recapture work. This is based upon the mean number of captures from each monthly cohort which are known to be alive in successive months, expressed as a percentage of the total number caught at time zero (when young first appeared in the population). It should be stressed that the disappearance of marked shrews from the population may not necessarily mean that they died; they may have left the area of study and not been recaptured in

Figure 2.6 *Survivorship curve for the common shrew* (Sorex araneus*) and the water shrew* (Neomys fodiens*) (after Churchfield, 1984a)*

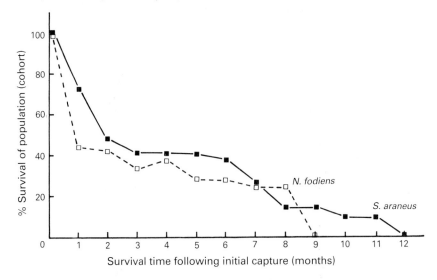

successive months. Hence, survival is equated with residency and not directly with mortality, which would be impossible to measure in such a population. Despite this, the data provide some idea of relative mean survival rates and life-expectancy at different times of life in shrews.

Both species showed high mortality and emigration during the first two months of life, with approximately 50 per cent of each shrew population being reduced in this way. *S. araneus* and *N. fodiens* then maintained a fairly steady survival from three to six months and three to eight months of age respectively, which is the overwintering period. As spring approaches and breeding commences, there is a decline in survival rate, which is pronounced in the case of *N. fodiens*, followed by a rapid die-off of ageing adults after breeding.

So life-expectancy varies from season to season in shrews, and they are most vulnerable at the juvenile stage and again at the breeding stage. In *S. araneus*, for example, it seems that only 40–50 per cent of individuals born into the population in summer survive to six months of age, and only 20–30 per cent survive to breed. Perhaps surprisingly, the time of greatest survival on a month by month basis is during the winter. It is common for the survival rate from month to month to exceed 90 per cent over the winter period, which is higher than at any other time of year (Churchfield, 1980a, 1984a).

CAUSES OF MORTALITY

Predation

There are a number of predators which will kill and eat shrews, but few of them take shrews as a major component of their diets. The most important predators are owls, but even they do not take as many shrews as they do other prey such as small rodents. For example, common shrews (*Sorex araneus*) constitute some 5 per cent by weight of the diets of tawny owls in deciduous woodland in England (Southern, 1954), and 6–13 per cent in the diets of barn owls (Buckley and Goldsmith, 1975). Pygmy shrews feature even less as prey of owls, constituting a mere 0.3 per cent of the total prey found in tawny owl pellets (Southern, 1954) and 5 per cent in barn owl pellets (Buckley and Goldsmith, 1975). Water shrews (*Neomys fodiens*), although larger, comprise only about 0.2 per cent of the total prey of these owls. The proportion of shrews in the diets of owls does vary somewhat from place to place, but they are always in a minority compared with rodent prey, whether it be in Britain, Germany or North America. The differences in the number of these three species taken as prey probably reflect differences in their population densities, and hence their availability. Kestrels and other raptors are also known to take shrews occasionally.

Shrews form a small fraction of the diets of weasels and stoats (Moors, 1975; Tapper, 1976), and are occasionally eaten by foxes and snakes. However, they are unpalatable to most mammalian predators, apparently because of their distasteful odour. Domestic cats, for example, will frequently catch and kill shrews but rarely, if ever, eat them. Shrews are smelly creatures: not only do the flank glands of mature males exude a strong odour, but there are other glands situated on the body of both males and females which, coupled with the production of smelly faeces, help to give shrews their distinctive odour. But their odour acts more as a means of communication in their social organisation than as a protection against predators. While the smell may repel some predators, it does not act as a very efficient deterrent in the case of domestic cats which persist in killing shrews, or predatory birds such as owls which have a poor sense of smell compared with their mammalian counterparts. However, some may argue that as shrews form such a small component of the diets of predators, their smell must act as a deterrent.

In his study of tawny owls over many years, Southern (1954) found that although shrews were taken regularly throughout the year there was a tendency for them to be caught more frequently in summer and autumn than at other times of year. It is at this time that shrews have the highest population densities, and there are many young, inexperienced shrews present which have recently left their parental nests. These shrews are probably most vulnerable to predation, since they are very active on the ground surface as they disperse to find new nest sites and establish their own home-ranges. However, the old adults may also be vulnerable to predators at this time, for they may be ousted from their home ranges and territories by fitter youngsters. Thus predation may be an important factor affecting the high mortality rates of these two age classes of shrews, although the actual loss from the population by this means is not known. Predation might be expected to increase in winter when the vegetation cover dies back and small mammals are more exposed as they forage on the ground surface. The fact that this does not happen in the case of shrews supports the evidence that they are less active on the ground surface at this time, and escape the attention of aerial hunters.

Parasites

Shrews are hosts for a great variety of parasites, both externally and internally. Fleas commonly occur amongst the fur, and may be specific to shrews. For example, *Doratopsylla dasycnema* and *Palaeopsylla soricis*, which frequent *Sorex araneus*, are found only on shrews. Mole fleas and several rodent fleas are also often found on shrews. Fur mites, nest mites and larval ticks are common. Many species of helminths (both tapeworms and flukes) are found in the stomach and alimentary canal of shrews. These are transmitted by the invertebrates such as beetles, slugs and snails on which shrews feed and which act as intermediate hosts for these parasites. Shrews tend to accumulate endoparasites as they age, and a build-up of 20 tapeworms in the gut is not uncommon. A summary of the parasites found to be associated with shrews in Britain gives some idea of shrews' importance as hosts and the great diversity of parasites they harbour (see Table 2.1).

It has been suggested by some that parasite loads may have a significant bearing on the mortality rates of shrews. Buckner (1969a), for example, found that the level of infection of *S. araneus* in Britain by *Porrocaecum talpae*, a very common spirurid-type nematode whose larvae are found encysted in the gut mesentery and connective tissue beneath the skin, could be correlated with a decline in the population numbers of shrews in autumn and winter. The percentage of individuals infected with *Porrocaecum* was greatest between November and January whereas between February and August the parasite was not recorded at all. He postulated that these parasites, which can reach 2mm in diameter when curled up *in situ*, may cause irritability and unrest in their host, making it more subject to predation and thereby contributing to a population decrease. The final hosts of these parasites are owls. Yudin (1962), working in the USSR, also found higher levels of infection by *Porrocaecum*, together with other helminth parasites, in overwintering shrews than in juvenile shrews, but he discovered no adverse pathological or behavioural effects attributable to the parasites.

In contrast, other workers have found that although there was a correlation between the levels of parasite infection of *S. araneus* and population density, declining populations of shrews were associated with decreasing rather than increasing levels of infection. Although the incidence and levels of infection by parasites does differ from year to year and geographically, workers in Wales (Lewis, 1968) and in Poland (Kisielewska, 1963) found that infection by helminth parasites such as the tapeworms *Choanotaenia crassiscolex* and *Staphylocystis scutigera* and the digenean fluke *Brachylaimus fulvum* was higher in summer juvenile and adult *S. araneus*, but significantly lower in overwintering animals. I found that although *Porrocaecum* was present at all times of year, it tended to decrease both in intensity and incidence in autumn when the shrew population was declining, and that infection by this and certain other helminths was lower in winter.

While shrews are usually the definitive hosts of their helminth parasites, they have one or two intermediate hosts—invertebrates, commonly gastropod molluscs or beetles, which are eaten by shrews. Hence, changing levels of infection in shrews may reflect differences in the availability

**Table 2.1   Parasites recorded from shrews in the British Isles
(●● = particularly common)**

	Sorex araneus	Sorex minutus	Neomys fodiens
Fleas			
Ctenophthalmus nobilis		●	●
Doratopsylla dasycnema	●●	●●	●●
Hystricopsylla talpae	●	●	
Palaeopsylla soricis	●●	●●	●●
Mites			
Euryparasitus emarginatus	●●	●●	●
Haemogamasus arvicolorum		●	
H. hirsutus		●	
H. horridus	●●		
Labidophorus soricis	●●		●
Ticks			
Ixodes ricinus	●●	●	
I. trianguliceps	●●	●	
Roundworms			
Capillaria incrassata	●		●
C. oesophagicola	●	●	
Longistriata depressa	●	●	
L. didas	●	●	
L. pseudodidas	●●	●●	
L. thomasi	●	●	
L. trus	●	●	
Parastrongyloides winchesi	●	●	
Porrocaecum talpae	●●	●	
Stammerinema soricis	●	●	●
Stephanskostrongylus soricis		●	
Tapeworms			
Choanotaenia crassiscolex	●●	●●	●
C. hepatica	●	●	
Coronacantha sp.			●
Hymenolepis schaldybini	●●	●●	
Neoskrjabinolepis singularis	●	●	●
Staphylocystis furcata	●	●	
S. scutigera	●●	●●	
Soricinia sp.	●	●	●
Triodontolepis bifurca			●
Vampirolepis sp.			●
Flukes			
Brachylaimus fulvum	●●	●●	●
Dicrocoelium soricis	●	●●	●
Opisthoglyphe exasperatum	●		
Acanthocephalans			
Centroryhnchus aluconis	●	●	●

of intermediate hosts. In winter, for instance, low levels of infection of shrews by *B. fulvum* and *C. crassiscolex* have been correlated with low population numbers of the gastropod *Zonitoides excavatus* and *Oxychilus helveticus* respectively, the major intermediate hosts of these parasites. This will in turn affect shrews for, if intermediate hosts are few in number, shrews are less likely to locate and eat them, and so infection is avoided.

It could be argued that the diminishing levels of infection by parasites which accompany the decreasing population sizes of shrews in autumn and winter are evidence that parasites kill the more heavily infected hosts, thereby causing the decline in population density of shrews. However, my studies have indicated that infection levels commence a decline several weeks prior to the reduction in population numbers of shrews, suggesting that parasitism is not a major factor in controlling population sizes and affecting the mortality of young. It is possible that high infection levels may contribute to the death of old adults in summer and autumn, but not juveniles or immatures, since these age classes generally have the lowest levels of parasitism.

Seasonal cycles of external parasites of shrews have also been found. Fleas and ticks are commonly seen on shrews, and a single animal may play host to three or four large ticks. While these may transmit viruses and other infections which could harm shrews (and about which little is known), they seem not to have any direct adverse effects. Detailed studies of the seasonal dynamics of the tick *Ixodes trianguliceps* by Randolph (1975) showed that larval tick infestations of *Sorex araneus* and *S. minutus* were highest in autumn and winter, but there were no lethal levels of infestation which might regulate the host populations.

Shrews such as *Sorex araneus* and *Neomys fodiens* are known to carry bacteria causing leptospirosis, tuberculosis and pneumonia; even rabies-related viruses have been discovered in tropical crocidurine shrews, yet there are no records of epidemics causing mass mortality as in the case of, say myxomatosis in rabbits. It seems unlikely, then, that parasitism is a major cause of mortality, although shrews are host to many different parasites and infestations often show seasonal cycles.

Old Age and Senescence

Following the sudden swell in population numbers in summer as young enter the population, there is a rapid decline in autumn in northern temperate regions. This has been termed the autumnal epidemic of shrews and it has two main causes. One, as suggested above, is the high mortality rate of young, inexperienced shrews which may be particularly vulnerable to predation as they achieve independence and disperse. The other is the death of the old adults still remaining after they have bred. Most adults seem to survive long enough to breed twice and many remain alive until late August and September, although there is a gradual loss of these animals through the summer as the survivorship curves suggest. There are various causes of death for these ageing shrews. One may simply be old age and senescence.

Shrews show quite distinct signs of ageing and senescence, as do many other mammals, despite their short life-spans. Externally, old shrews may appear fit, sleek and healthy, except for the presence of increasing numbers of grey hairs, a grizzled appearance on the head and rump, and the loss of the bristly hairs on the tail making it progressively more naked—all features which accompany old age. However, old shrews may not undergo a full autumn moult to provide them with a thick winter coat. Often moults are incomplete, and some individuals can become quite bald. Occasionally they show other signs of wear and tear: toes, feet and even whole limbs may be red and greatly swollen for no obvious reason. Toes may be missing, and one wild shrew was found to be running around foraging and breeding for at least two months minus an entire front foot. Approximately 10 per cent of the population of *Sorex araneus* has been found to have injuries of one sort of another, but whether these arise from disease, infection, accident or fights with other shrews is not known. One wild shrew caught during the breeding season was found to be badly bitten around the head and face, injuries which could have been sustained in a fight or during a lucky escape from a predator. It soon disappeared from the population being studied, and was presumed to have died from its injuries. Captive shrews sometimes develop tumours and carcinomas both externally and internally, but these are seldom seen in wild populations and so may result from dietary deficiencies.

A major sign of ageing is the condition of the teeth, and the rate of tooth wear has been used to estimate the age of shrews. When studying the population dynamics of species it is important to be able to determine the ages of the component individuals and assign them to particular age classes, since the age structure of a population has an important bearing on its rate of growth and the survival and life-expectancy of its members. The body weight of individuals provides some guide to their relative ages, but it is only a crude one and often not very useful. For example, many species may achieve near adult weight in youth and undergo little further increase throughout the remainder of their lives, which may extend for several years. In the case of shrews, which are very short-lived, the problem is similar. By the time juvenile shrews have left their parental nest they weigh nearly as much as the adults and, more importantly, they do not continue to grow at an even rate through ensuing months. Many species in temperate regions lose weight during the autumn and winter and then undergo a spurt of growth to sexual maturity in spring. So while it may be possible to assign shrews to an age class (juvenile, sub-adult and mature adult) on the basis of body weight, it is not possible to age them accurately even to within a month. A sub-adult found in November could have been born in May, June, July or August. In some species, such as *Crocidura russula*, it is difficult even to distinguish adults from juveniles. So the rate of tooth wear provides a more accurate method of ageing shrews.

The teeth of shrews do not grow throughout life, as do the incisors of rodents, but they wear down and are not replaced. The amount of abrasion ranges from virtually unworn in very young animals to teeth worn down almost to the gum line in very old shrews. Tooth wear results from the continual abrasion during feeding and the nature of the diet. Most prey comprise invertebrates with hard, chitinous exoskeletons which have to be chewed up and, for those species feeding extensively on earthworms, the soil particles in the intestine of these prey provide a major source of abrasion. While the rate of wear may depend upon the diet, the soil type and other environmental factors, the amount of wear can be correlated with the age of the shrew, in weeks.

With red-toothed shrews including species of *Sorex* and *Neomys*, a progressive loss of the red-stained enamel of the teeth is a good indication of tooth wear and age. The technique employed most often for quantifying tooth wear in shrews is to examine selected teeth and measure their height. Crowcroft (1956) used the second tooth from the front (the canine) on each side of the lower jaw since this received as much wear as any other tooth and was the easiest to measure. Using a microscope, he measured the height of each tooth from the edge of its socket to its tip, and got a useful correlation between the wearing down and the relative age of the shrew. A similar technique has been used in white-toothed shrews. For example Jeanmaire-Besançon (1986) estimated the monthly rate of dental abrasion at 32μm. For shrews, as for other small mammals, this technique is impossible to perform on living populations, which is a major drawback.

Shrews are not long-lived animals, but it seems unlikely that wild shrews die simply because they have outlived their physiological life-span; their lives can be prolonged, sometimes by several months, in captivity. It seems that death may result from a combination of old age and sociological factors.

Sociological Factors

Sociological factors may influence mortality rates of old and young shrews alike. In summer and autumn, when population densities are at their highest, there must be numerous contacts between individuals as they search for food and territories. This may well give rise to stress as meetings between shrews invariably entail agonistic reactions by way of postures and vocal signals, and sometimes true aggression results when fights occur. Shrews do fight, even when unconfined in the wild, but it is unlikely that they actually kill each other in the process. But, bearing in mind the narrow line between life and death for a shrew, any harassment or interference which prevents feeding for a length of time could be critical.

Since most adults are dead before the advent of harsh winter conditions, it seems unlikely that adverse climatic factors hasten their demise. However, it may be that climatic factors augmented by sociological factors such as competition with younger shrews for food and nesting sites, and resulting agonistic behaviour at this time of high population density, accelerates the death of aged adults. There is some evidence that young shrews are socially dominant to old shrews and are able to hold territories against them. Once the old shrews have been ousted from their home-ranges or territories they may be more prone to predation and food shortage.

SEASONAL WEIGHT CHANGES
AND OVERWINTERING

Winter Weight Loss

Most shrews in temperate regions, including common and pygmy shrews (*Sorex araneus* and *S. minutus*) undergo seasonal changes in weight with a marked decrease during autumn and winter. Young common shrews, for example, achieve body weights of 7–8g by September–October but, as the winter approaches, they decrease to as little as 5.5g in Britain. Their body weights reach a minimum between December and February and then there is a great spurt of growth in March–April as they put on weight and grow to sexual maturity. By the summer these fully grown shrews weigh 10–12g, sometimes more.

The phenomenon of winter weight loss is widespread, and it has been known for many years. It was first recorded by Adams in 1912 as he investigated the natural history of shrews in England. It has subsequently been reported in a number of different species including *Sorex araneus*, *S. minutus* and *Neomys fodiens* in Europe, and *S. vagrans* and *S. fumeus* in North America. It is particularly well documented in *S. araneus* in Britain, the Netherlands, Poland, Finland and Germany. It seems to be chiefly, if not exclusively, a feature of northern temperate soricines, primarily members of the genera *Sorex* and *Neomys*, for significant winter weight losses have not been recorded in crocidurines whose geographical ranges overlap those of *Sorex* species in some northerly regions. *C. leucodon* undergoes only a slight decrease in winter, and *C. russula* and *C. suaveolens* merely show a slackening in the rate of growth, with body weight continuing to increase slowly.

It also seems that there is a geographical basis to the phenomenon. In Britain, for example, the winter weight loss amounts to about 27 per cent in common shrews (Churchfield, 1981), 37 per cent in Poland (Pucek, 1965), and in Finland it can reach a massive 45 per cent (Hyvarinen, 1969). Winter conditions in different geographical areas seem to affect the changes which occur: they are greater in northern and eastern Europe where climatic effects are more severe than in Britain with its milder climate.

The phenomenon is seen on both an individual and a population basis. Some examples of this remarkable weight change are shown in Figure 2.7 where the weights of known, marked individuals of *Sorex araneus* in the wild have been recorded over successive months in a scrub-grassland study area in central England. Figure 2.10 shows the effects of these weight changes on the profile of the whole population of these shrews.

Causes of Weight Loss

What is more remarkable about the winter weight loss is the way in which it is manifested, and this has been the subject of considerable attention by researchers. It seems that the loss in weight is due to a combination of morphological and physiological changes, including a decrease in the dimensions of the skeleton and certain internal organs, leading to a slight reduction in overall body size as well as weight.

Figure 2.7 Examples of winter weight changes in five wild common shrews (Sorex araneus) *in central England*

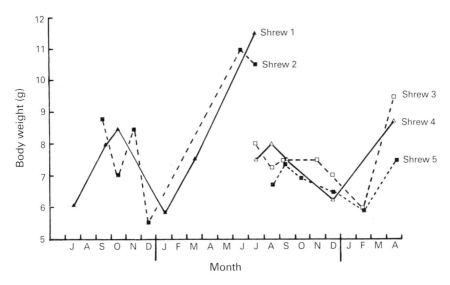

It was Dehnel (1949) who discovered that the reduction in body size of *Sorex araneus* in autumn in Poland was accompanied by a shrinkage in the size of the skull, especially in the reduction of the height or depth of the brain-case. The same feature was also then found in *S. minutus* and *Neomys fodiens*. This was later confirmed by Crowcroft and Ingles (1959) for *S. araneus* in Britain. The skull decreases in size between October and January, increases thereafter until July and then decreases once more towards the end of the shrew's life (see Figure 2.8). Since Dehnel first described the way in which such morphological changes contribute to the winter size decrease in shrews, the phenomenon has been known as the Dehnel effect. This is an apt tribute to the Polish zoologist who was a pioneer in the study of shrews. This phenomenon is characteristic of *Sorex* species and, in Europe at least, northern and eastern populations undergo a greater decrease than do southern and western ones.

Other workers have gone on to discover how these skeletal changes actually occur. Hyvarinen (1969) found that the decrease in the height of the skull is caused by resorption of bone at the edges of the parietal and occipital bones. New bone is then formed here in the spring. A decrease in overall body length is caused by changes within the intervertebral discs of the spinal column and resorption of cartilage in the discs in winter. The activity of alkaline phosphatase is very low during the resorption period but high during the growth period, and this is partly connected with the strong secretory activity of the parathyroid gland in autumn and early winter. The parathyroid gland has an important role in maintaining calcium levels in the body. The increase in body size in spring is associated with a sudden increase in somatotropin secretion in the pituitary gland which stimulates growth. Thus, the seasonal changes in the body have a strong hormonal basis.

Not only do the skeletons of young shrews undergo changes in winter,

Figure 2.8 Seasonal changes in the depth of the brain-case of the common shrew (Sorex araneus) in Britain. Vertical lines denote 95 per cent confidence limits (after Crowcroft and Ingles, 1959)

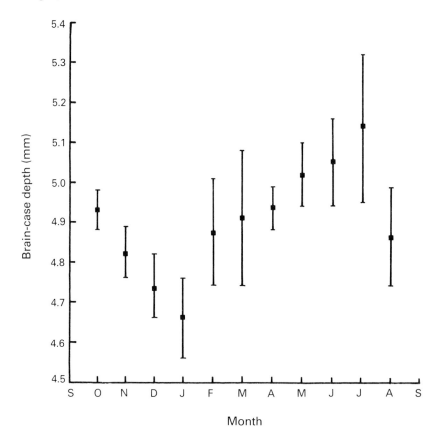

but so do many of the internal organs. A decrease in the size of the brain itself, plus the liver, kidneys, spleen, adrenal glands and thymus have been found (Pucek, 1965). The seasonal cycle in the size of some of the internal organs can be seen in Figure 2.9. These organs then increase in size again in the spring. Even the reproductive organs undergo a change: although they are small and underdeveloped in immature shrews in summer, autumn and winter, nevertheless the testes of the males and the ovaries of the females are slightly smaller in December–January than they were shortly after the young shrews left the parental nest. Some regression in these reproductive organs is also found in old adults as the summer and autumn progresses, following breeding.

Cycles in body water content and in fat content are also found. Myrcha (1969) and Pucek (1970) explained the seasonal changes in body weight, particularly of the internal organs, by tissue dehydration in autumn and winter, behind which were morpho-physiological changes in the endocrine glands and variations in metabolism. But could tissue de-hydration account for such large reductions in body weight—in the order of 30 per cent? The seasonal cycle in body water content certainly reflects

Figure 2.9 Seasonal changes in the weight of internal organs of the common shrew (Sorex araneus) *expressed as a percentage increase or decrease of monthly averages. Average for sub-adults in June taken as 100 per cent (after Pucek, 1970)*

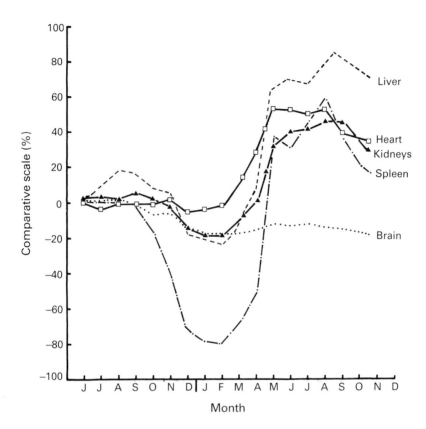

the changes in body weight, as can be seen in Figure 2.10. However, I found that the change in water content between juvenile *Sorex araneus* in summer and sub-adults in winter in Britain was only in the order of 3–4 per cent, with a maximum of 6.25 per cent, which accounted for less than a quarter of the weight loss shown by shrews (Churchfield, 1981). So, although differences in water content may contribute to the weight changes of shrews, they are not the main factor involved.

Changes in body fat content show a rather different pattern. While there is considerable variation in the seasonal fat levels of shrews from one population to another and from one year to another, the general pattern is for fat content to be lowest in summer and highest in autumn and winter (Pucek, 1965; Myrcha, 1969; Churchfield, 1981). However, the amount of fat stored by wild shrews is quite small, only amounting to 4.5–7.5 per cent in *S. araneus*, and the magnitude of these seasonal changes seems insignificant, with a maximum of only 2–3 per cent of the body weight in Britain and Poland. Fat levels differ from area to area: for example, both Myrcha in Poland and I in England found levels were consistently higher in the more northerly study areas than further south.

Figure 2.10 Seasonal changes in the mean body weight and percentage water content of the common shrew (Sorex araneus) *in central England. Vertical lines denote standard errors (after Churchfield, 1981)*

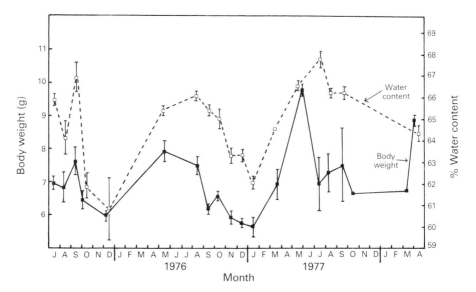

Nevertheless, the importance of fat levels to the overwintering success of these shrews would appear to be negligible, especially compared with values for captive shrews. Shrews can accumulate fat rapidly, within one or two months of captivity, and to a massive 30 per cent or more of their total body weight. It has been suggested that wild shrews accumulate fat in autumn for use in winter, when energy expense is high because of low temperatures, and in spring when there is an increase in growth and activity with the onset of the breeding season. However it is unlikely that these very small fat reserves are of much use as energy stores, although they may aid thermal insulation. It seems that brown adipose tissue (BAT) is the only form of fat reserve found in shrews under natural conditions, and it is concentrated in the shoulder areas as sub-scapular deposits. BAT is known for its prominence in young mammals, including human babies, and is thought to be important in heat production and metabolism, assisting the survival of newly born animals. It also has an important role as a fat store in hibernating mammals.

Since maximum fat levels are found in the depths of winter, the fat reserves of shrews do not seem to be utilised at this time of year as they are in summer when values are lower. Low summer fat levels may result from the drain on reserves caused by the increased activity associated with breeding. In winter, reduced activity and increased food abundance could explain the higher fat values, and there is some evidence to support this.

The fact that captive shrews readily lay down fat reserves while wild ones do not suggests that conditions for wild shrews are never conducive to fat storage, and that there is some feature of their ecology or physiology which does not permit it. Food supply is ruled out as the limiting factor by the failure of captive shrews kept in outdoor enclosures, under ambient conditions but with abundant food, to undergo fat deposition. Experiments suggest that temperature may be an important factor, but even this seems unlikely for captive shrews indoors at 20°C will accumulate fat whereas wild ones in summer will not. On the other hand, fat deposition in captive shrews may result from the constancy of the environment coupled with low levels of activity and lack of competition and stress from other shrews.

It seems, then, that changes in water and fat content, and the other morphological and anatomical changes which have been recorded, are not of sufficient magnitude to account fully for the winter weight loss of shrews, although they do all contribute to it. It is possible that the major part of the weight loss is caused by a reduction in overall protein content, which results in a decrease in musculature in a similar way to the seasonal changes in temperate birds such as starlings.

Reasons for the Winter Weight Loss

Several reasons for the loss in body weight and dimensions have been suggested. Food shortage coupled with the physiological strain of moulting and a drop in ambient temperature in autumn has been proposed (Niethammer, 1956). An increase in food supply and a rise in temperature supposedly permits a gain in body weight in the spring, despite the so-called strain of a spring moult. But there is no evidence of a spring increase in food supply to account for the rapid weight gain.

The role of food supply and climate in the seasonal weight changes of shrews is equivocal, especially since captive shrews maintained in outdoor enclosures at ambient temperature but with an abundant food supply will undergo a decrease in weight from September–October to December which parallels that exhibited by wild shrews (Churchfield, 1979). Figure 2.11 shows the weight changes of captive common shrews under winter conditions. The weight decreases accompanied a decline in ambient temperature, but although minimum temperature continued to fall to as little as –6°C, no further decrease in weight occurred. In spring, when minimum temperatures were no more than 1°C, body weight increased towards maturity. The exceptionally cold spring did not affect the timing of maturity in these shrews which pursued a pattern of growth similar to that in wild shrews. Captive white-toothed shrews (*Crocidura russula* and *C. suaveolens*) apparently show no winter decrease under these conditions; it is not known if they do so in the wild.

That temperature is not the prime limiting factor in growth is shown by the fact that wild shrews, and captive ones maintained outside, will increase in weight in spring despite the prolongation of cold weather. It may also be concluded that seasonal weight changes cannot be attributed to the availability of food resources. In Britain at least, there is no evidence of a shortage in biomass or numbers of prey available in winter, and certainly there is not a sudden increase in food supply in spring at the time when shrews exhibit a spurt of growth (Churchfield, 1982a). The pattern of growth and the abundance of prey in different seasons in Britain are shown in Figure 2.12. If anything, there are more ground- and soil-dwelling invertebrates around in winter than in

Figure 2.11 Winter weight loss of three captive common shrews (Sorex araneus) *kept in outdoor enclosures, together with changes in daylength and minimum ambient temperature*

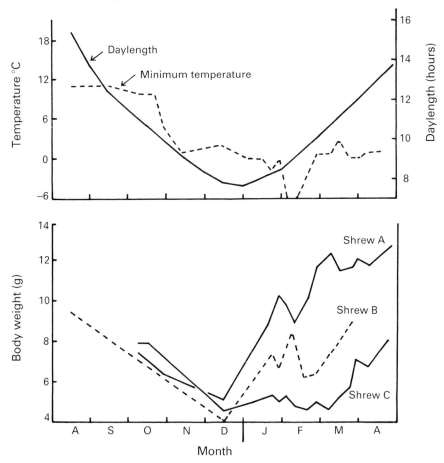

Month

summer. It is possible that low temperatures inhibit feeding by producing a form of hypothermia which slows down the activity of the shrew, making it physically incapable of either catching or eating food quickly enough to maintain its body weight. However, there is no evidence of a significant reduction in metabolic rate or body temperature which could cause a decrease in assimilation rate and a digestive bottleneck. A reduction in activity produced by low temperatures could create a feeding bottleneck but, again, this seems unlikely when considering the rapid spring increase in growth and activity which occurs while conditions are still cold.

The relationship between seasonal weight changes and daylength is interesting. Captive *Sorex* maintained at constant temperatures of about 20°C and a constant daylength of twelve hours show large variations in weight between individuals. Some exhibit a winter decrease under such conditions, but most do not. Instead, they grow gradually to maturity and reach breeding condition in late winter or spring. A major problem with

*Figure 2.12 Mean monthly body weights of cohorts of common shrews (*Sorex araneus*)
together with seasonal biomass and numbers of invertebrate prey (after Churchfield, 1982a)*

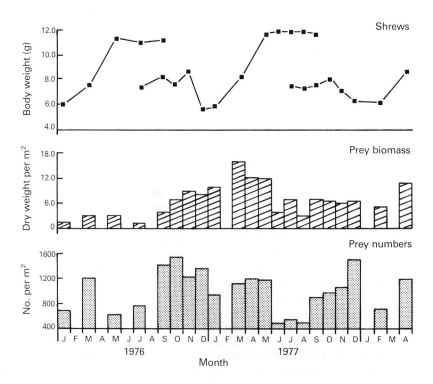

studying shrews is that their body weights are highly variable and may
change quite markedly (±0.5g) from day to day. The relative fullness
of the stomach and gut influences such changes in body weight, depending upon the timing of the last meal.

A close correlation between daylength and the seasonal body weight
changes of *Sorex* kept outdoors suggests that the photoperiod may have
an important influence (Figure 2.11). Cues for changes in the body weight
of temperate soricine shrews may be provided by both daylength and
temperature changes acting on the pituitary and parathyroid glands
which control hormone levels, restricting growth in autumn and then
promoting it in spring. The detailed work of Hyvarinen (1967, 1969) has
indicated the importance of the endocrine system in the seasonal weight
changes of common shrews, particularly the thyroid and pituitary glands.
He found a clear seasonal and age-related variation in the activity of the
thyroid gland as suggested by the height of the constituent follicular
epithelium and the size of the nuclei of this gland in shrews in Finland, as
can be seen in Figure 2.13. These were very small in juvenile shrews in
winter but increased towards autumn. Activity of the thyroid remained
high in November and December, but decreased markedly in January,
February and March before a rapid increase to its highest level in spring.
It then declined slightly as the animals aged. The strong thyroid activity
in autumn was associated with moulting, and also with acclimation to

cooler weather. The decrease in midwinter was assumed to be correlated with severe winter conditions when the shrew reduces its activities, including its metabolism, and uses its energy as economically as possible. This coincided with a reduction in the activity of the pituitary gland in winter, as indicated by its size (see Figure 2.14), followed by a rapid increase in spring as the shrew grows to sexual maturity. Further research is needed to investigate the precise conditions under which these changes occur before the mechanisms can be fully explained.

*Figure 2.13 Seasonal changes in the size of the nuclei of the follicular epithelium cells of the thyroid gland of the common shrew (*Sorex araneus*). Vertical lines denote standard errors (after Hyvarinen, 1969)*

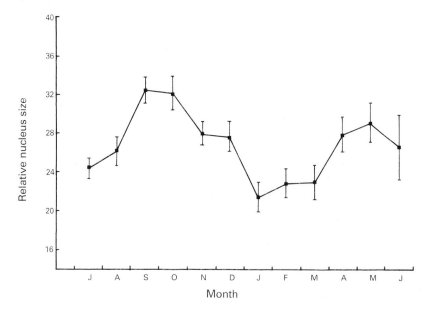

*Figure 2.14 Seasonal changes in the cross-section area of the anterior lobe of the pituitary gland of the common shrew (*Sorex araneus*) (after Hyvarinen, 1967)*

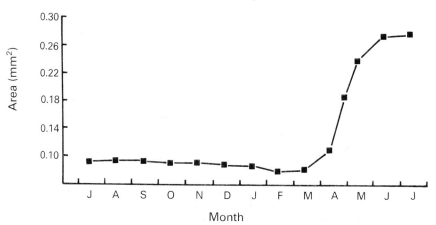

Very little is known about seasonal changes in tropical and sub-tropical shrews. It seems that they mostly grow steadily from youth to sexual maturity. In these species, seasonal patterns of growth are likely to be affected by the periodicity of alternating dry and wet seasons which greatly affect the food supply, particularly in the drier habitats such as grasslands. Dry seasons may well exert a check on the growth of these shrews. Some crocidurines are able to become torpid for short periods, and they may lose weight at such times. This will be discussed in more detail in Chapter 6.

Advantages of a Decrease in Body Size

It has been proposed that the shrews' loss in size and weight in winter may be a strategy to reduce food requirements. A small animal eats less food than a large one and so would need to spend less time foraging in cold winter conditions and could therefore conserve its energy. In this context it is interesting to note that some of the smallest species of shrew have ranges which extend far north towards the Arctic Circle. These include *Sorex minutus*, *S. minutissimus*, *S. arcticus* and *S. caecutiens*. For animals which are unable to hibernate, this could be a very useful strategy, but it is counterbalanced by the greater heat loss in small as compared with large bodies. Small bodies, with their relatively high surface area to volume ratios, lose heat more rapidly than larger ones, and this is a serious disadvantage in a cold climate. Thus it is not an accident that polar representatives of many mammalian families are larger than their temperate or tropical relatives, a phenomenon first described in 1847 by Bergmann, who observed that the dimensions of warm-blooded animals increased towards the north. However, this is clearly not the case in shrews. Indeed, there are many species which are now known not to conform to Bergmann's rule, but the reason for this is not well understood.

There must be some overriding advantage in remaining small, despite the thermoregulatory penalties, and as I have said, the explanation may lie in reduced food requirements. This will be discussed in more detail in Chapter 6.

MOULTING

Moulting is a regular seasonal activity of shrews. In northern temperate regions, for instance, they moult twice a year, in autumn and again in spring. In September–October the common shrew (*Sorex araneus*) moults from its juvenile pelage of short summer hairs which are only some 3.5mm in length, to a thicker and longer winter coat with hairs 6–7mm in length. This moult commences at the rump and progresses forwards to the head in this species, with the dorsal surface moulting before the ventral surface. The sequence of the autumn moult is show in Figure 2.15. In spring, this thick winter coat is exchanged for a sleek summer one as the shrew undergoes two moults in quick succession between March and May, with females completing their moults rather earlier than males. But this time the moults commence at the head and proceed

*Figure 2.15 Stages in the autumn moult of the common shrew (*Sorex araneus*)*

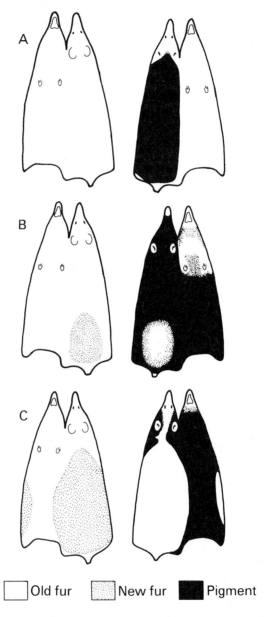

Old fur   New fur   Pigment

posteriorly towards the rump. Not only does the length of the hairs differ from summer to winter but so too does the thickness of the skin. It is about 50 per cent thinner in overwintering sub-adults than in summer juveniles or spring/summer adults. The actual process of the moult lasts only 2–5 days. Pygmy shrews (*Sorex minutus*) undergo the same regime of moults, but in this species the moults proceed in the opposite directions to those of the common shrew.

3 Social organisation

HOME RANGES AND SPATIAL DISTRIBUTION

The majority of shrews are solitary and asocial for most of their lives, and avoid contact with their fellows. As a juvenile, each shrew establishes its own home range, a small area in which it nests, forages and explores, and where it remains for most of its life. Individuals, once they have established their home range, can be captured and recaptured in the same small area month after month, indicating that there is a high degree of site-attachment. Shrews therefore maintain characteristic patterns of spatial distribution with respect to their own activities and those of their fellows. At certain times of year there is evidence that these home ranges become much more clearly defined with respect to conspecifics. Overlap between them is minimal and they show distinct signs of territoriality.

DISPERSAL AND MOVEMENT PATTERNS

As soon as young shrews are weaned, they begin to disperse from their natal areas, and are encouraged to do so by the increasing hostility expressed towards them by the mother and by their siblings. If the mother is to breed again, and use the same home range and nest site to bring up a second litter, it is particularly important that these young shrews disperse quickly to avoid competition for resources (such as food) with the mother and the next cohort of juveniles. It is also important that they establish themselves quickly in their own home range, with which they can become familiar, and gain the necessary information about good foraging and nesting sites and avoid predation.

With large numbers of juveniles being born, populations of northern temperate shrews are in great flux during early and midsummer, and there is much movement of the young shrews as they search out suitable home ranges. The breeding females show considerable site-dependency and quite fixed home ranges, but the adult males wander extensively in search of oestrus females with which to mate. So at this time of year when the shrews are breeding, there is much overlap of home ranges and invasion of old adults' territories by juveniles.

At first there is room for the young shrews to take over home ranges close to the natal area, replacing old adults which have died or mature males which have gone off in search of females, and so there is no need for juveniles to disperse more than a few metres from their area of upbringing. As the population grows and more youngsters appear, vacant home ranges are more difficult to locate and the latest litters of young have to disperse further and further from the natal area, making them vulnerable to predation and accident. Dispersing shrews often have lower body weights than resident juveniles and it seems that those shrews which disperse furthest from their natal area are the smaller individuals. It is probably that the first-born young take over the nearest, vacant home ranges, gain weight quickly and successfully deter invasion by later, smaller juveniles so that these are forced to go further afield, even into sub-optimal habitats. There must be a distinct advantage for adult females, who invest heavily in the birth and rearing of offspring, to breed as early as they can in order for their young to take over the nearest, vacant and best home ranges.

Some species of shrew seem to have greater dispersal abilities or inclinations than others. Water shrews (*Neomys fodiens*), for example, disperse very widely. Young animals, in particular, can be found far from their natal area, often hundreds of metres, even several kilometres, from water. In population studies the recapture rate of this species is low relative to other species such as *Sorex araneus* or *S. minutus*, suggesting that it is generally a nomadic species. Individuals tend to move into an area, remain for a few days and then move on. They are also capable of prolonged daily excursions. Shillito (1960) found that water shrews would move up to 162m in one day, while the maxima (and the exceptions) for common and pygmy shrews were 144m and 60m respectively. They seem to have a system of shifting home ranges, particularly amongst juveniles. They leave the streamsides where space becomes limited as the population increases during the summer breeding season. They travel through the countryside until other suitable habitats are reached. This explains their sudden appearance and disappearance in hedgerows, woods or grasslands far from streams and rivers.

Pygmy shrews also seem to disperse relatively long distances. Despite their small size, individuals sometimes travel 80m or more as juveniles.

Compared with these two species, the common shrew is remarkably sedentary, most remaining within a few metres of their natal area. However, some individuals wander considerably. Juveniles have been found to disperse up to 119m, but they probably do go much further afield. Wandering males have been found up to 144m from their original home range area, travelling up to 135m in one day.

Because of the limitations of live-trapping studies we have little idea how far shrews may travel as they disperse or search for mates, but one male *Sorex cinereus* was recaptured as an adult approximately 800m from its last capture point as an immature the previous autumn.

The predisposition of some species to disperse while others do not has important biogeographical implications. *Sorex minutus* clearly has a better dispersal ability than *S. araneus* because it occurs in a variety of far-flung places which have not been reached by *S. araneus*. These include many of the Outer and some of the Inner Hebrides, the Orkneys and the Shetland Isles. It does not even share the Isle of Man or Ireland with the common shrew. It seems that this tiny shrew dispersed more quickly than the common shrew after the last glaciation, and that it reached and colonised these areas, which were afterwards cut off by the rising sea as the ice melted. This has been attributed to the more elastic habitat and food requirements of the smaller shrew. In contrast to *S. minutus*, the common shrew relies heavily on earthworms as a food source and shows an interesting parallel with the European mole in its distribution. Moles cannot live in permanently frozen soil and so could only colonise areas exposed by the retreating glaciers some time after the soils had warmed up. Neither species thrives in acid, peaty moorlands where there is a paucity of earthworms, whereas *S. minutus*, which feeds on small arthropods, has no such restrictions to hamper its dispersal.

There are also genetic implications for the differences in dispersal between species, highlighted by *S. araneus*. As we have seen, this species exhibits chromosome polymorphism, and a number of distinct races have been identified in parts of Britain and Europe. These may have arisen because populations did not disperse or mix. Thus populations became locally isolated and evolved in their own characteristic way. In contrast, *S. minutus*, with its wider dispersal abilities has not been found to exhibit this phenomenon.

Once breeding is completed and most of the old, mature adults have died off, the amount of movement within the population declines and the remaining individuals settle into an established and characteristic pattern of spatial distribution. In winter, live-trapping data show that activity and movement of common shrews on the ground is much reduced (see Table 3.1). The changes in movement patterns of shrews through the seasons are shown in Figure 3.1.

Table 3.1 Movement of common shrews in summer and winter

	Mean total distance moved within trapping period (m)	Mean distance moved between captures (m)
March–September	35	15
October–February	17	9

As autumn progresses and winter comes, the home ranges of the young shrews develop distinct boundaries and individuals organise themselves into territories.

TERRITORIALITY IN SHREWS

The definition of territoriality is the defence of an area from which conspecifics are excluded (Burt, 1943; Ewer, 1968). A territory comprises a distinct area with well-delineated boundaries which is defended against intruders of the same species. In many large mammals which exhibit territoriality, careful observation of individuals in the wild can reveal

*Figure 3.1 Changes in movement patterns of common shrews (*Sorex araneus*) in different seasons (after Churchfield, 1980a)*

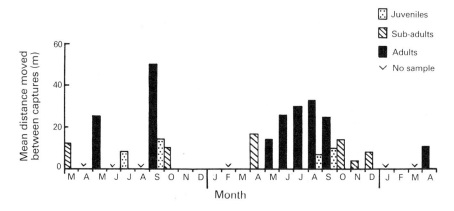

how their territories are established, patrolled and defended. But in small mammals such as shrews, which remain hidden amongst dense vegetation, it is difficult, if not impossible, to make good observations of them in their natural state. Evidence of their territoriality must come indirectly from mark-recapture information gathered during live-trapping studies, from close-range radio-tracking or the use of radio-active tagging which reveals the basic pattern of distribution and area of activity of selected individuals. Such evidence can then be supplemented by observation of captive individuals and their reactions to conspecifics.

Perhaps the best evidence for intraspecific territorial behaviour in shrews comes from the extensive work on common and pygmy shrews (*Sorex araneus* and *S. minutus*) by Michielsen in 1966. She studied populations of these shrews in an area of dune scrub-grassland in the Netherlands by an ambitious programme of live-trapping by means of pitfalls and Longworth traps. She marked each shrew with a small silver leg-ring stamped with a personal identification number, and plotted the pattern of distribution of each individual month by month by keeping careful records of the trap points at which they were captured.

Michielsen's live-trapping programme revealed that the home ranges of immature shrews of each species were almost completely non-over-lapping in autumn and winter, from which she concluded that both sexes were strictly territorial at this time of year. Individual home ranges had clearly delineated boundaries, often coinciding with distinct vegetational changes or topographical features, and remained in the same positions throughout the autumn and winter. In most cases the home ranges were adjacent to each other but were mutually exclusive, forming a jigsaw of tenanted areas (see Figure 3.2). Where overlap did occur it was at the periphery of a home range and, she discovered, it was often a sign that one home range resident had died or emigrated and the neighbour was busily annexing part of the newly vacated area. Vacant areas were quickly filled by neighbouring shrews, and so the shape and size of each territory was determined by the residents of adjacent territories.

Michielsen encountered sporadic one-off captures of shrews inside their

*Figure 3.2 Arrangement of individual home ranges of immature common shrews
(Sorex araneus) and pygmy shrews (S. minutus) in autumn/winter revealed by live-trapping.
Solid lines represent male ranges, dotted lines female ranges (after Michielsen, 1966)*

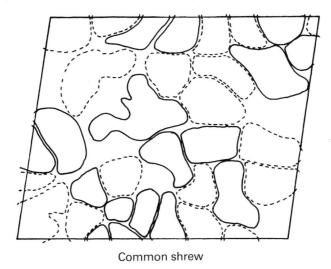

Common shrew

Pygmy shrew

neighbour's territory, followed by capture back in the home territory, as
though a quick sortie was being made to test the ownership of a
neighbour's territory. This territorial trespassing was more frequent in
S. minutus than *S. araneus*, making it more difficult to define the borders
of individual ranges in this species.

    She concluded that aggression must play an important part in main-
taining territories in both these species, and that the boundaries must be
fixed primarily by fighting. Since invasion of vacant territories occurred

very rapidly, she deduced that border disputes must be a regular feature of the daily lives of these shrews. Rapid takeover of vacant territories also led her to conclude that intraspecific population pressure was high. Those individuals which did shift their home ranges to different areas were those which had previously occupied only small territories.

There was no difference in territory size between the sexes in either shrew but there was a significant difference between the species. Though considerably smaller in body size, the mean territory size of *S. minutus* was an average of 2.2 times larger than that of *S. araneus* in winter and 1.5 times larger in late summer. Territories ranged from 370 to 630m^2 in the common shrew and from 530 to 1,860m^2 in the pygmy shrew (see Figure 3.3). The sizes of territories did differ considerably between individuals and seemed to vary with habitat: territories were smaller in scrubby, partly wooded areas than in more open grassy areas.

Figure 3.3 Home range sizes of common shrews (Sorex araneus) and pygmy shrews (S. minutus) in different months. Vertical lines denote ±1 standard error (after Michielsen, 1966)

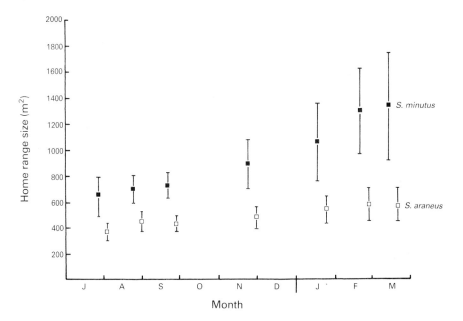

Territory size in both species tended to increase with time during autumn and winter as annexation of vacant areas occurred. The mortality rate of *S. minutus* was higher than that of *S. araneus*, and so the rate of increase in territory size amongst survivors was rather greater in this species.

As these two shrews mature in the spring, their spatial distribution changes and territoriality begins to break down. Both sexes extend their activities well beyond the confines of their winter territories, males more so than females. This leads to considerable overlap of home ranges. Although females remain in much the same area during the breeding season, males become great wanderers, crossing the home ranges of neighbouring males in their search for oestrous females. Several males at a time may be attracted to a single female's range. The mature males may return periodically to their original home ranges, but strict territoriality is abandoned as the rut proceeds.

A similar system of territoriality in winter but abandonment in spring and summer was found in two *Sorex* species in North America by Hawes (1977), suggesting that this may be a common pattern amongst northern temperate shrews. Hawes carried out a similar programme of live-trapping and mark-recapture on *S. vagrans* and *S. obscurus* which coexisted in an area of mixed forest in British Columbia. Like Michielsen, she also constructed maps of home ranges of marked individuals based on their positions of capture from month to month. She found that by autumn the young shrews had established virtually non-overlapping, contiguous home ranges or territories with respect to others of the same species. The little overlap that did occur involved individuals which did not remain in the area until the breeding season. She noted that most mortality and emigration occurred during the summer when territories were being established, but that nearly all shrews holding territories in autumn survived the winter to reach sexual maturity in spring.

These shrews increased their home ranges greatly in the spring with the onset of sexual maturity, although the position of each individual's range remained in approximately the same location as the winter territory. Both species had home ranges or territories of a similar size (see Table 3.2). The home ranges in spring and summer overlapped considerably, both in mature and young shrews.

Table 3.2   Home range sizes of *Sorex vagrans* and *Sorex obscurus* in different seasons (after Hawes, 1977)

	Mean home range size m^2	
	IMMATURE autumn/winter	MATURE spring/summer
Sorex vagrans	1,039	3,258
Sorex obscurus	1,227	4,020

Although there was no difference in territory size between the sexes in immature shrews of either species, the home ranges of mature males were about twice as large as those of mature females (see Table 3.3).

Table 3.3   Home range size of male and female shrews in spring and summer (after Hawes, 1977)

	Mean home range size m^2	
	MATURE MALES	MATURE FEMALES
Sorex vagrans	4,343	2,233
Sorex obscurus	5,978	2,226

During the breeding season Hawes identified three types of individual in the population: natives, immigrants and transients. Natives were well-established individuals which had remained in the same location and home range since youth. Most of these were females, and nearly all females which bred successfully were natives. Immigrants were unmarked individuals which entered the population when sexually mature and remained for at least 30 days in the study area. Transients also arrived in a sexually mature state but stayed for less than 30 days. Immigrants and transients were usually males, presumably searching for mates.

Studies of the short-tailed shrew, *Blarina brevicauda*, suggest that a similar but rather more rigid pattern of territorial behaviour occurs on a seasonal basis. Overlap of home ranges is minimal amongst immature shrews in winter. In summer, home ranges of members of the same sex do not overlap but those of opposite sexes may do so, implying again that there is considerable movement in the search for mates.

Information about home ranges and territoriality of shrews in other geographical regions is scarce, since few population studies have been carried out on these animals in tropical and sub-tropical areas. However, evidence does suggest that the mouse shrew, *Myosorex varius*, found in grassy habitats in southern Africa, has a similar system of non-over-lapping home ranges or territories when immature, but that overlap increases when they became mature, especially between male and female ranges. Evidence of territoriality in the African musk shrew, *Crocidura flavescens*, has also been found. These shrews are solitary for most of their time, and one study revealed a clear-cut boundary between the home ranges of two wild males, although male and female ranges partially overlapped.

So, shrew populations remain fairly static in autumn and winter with respect to spatial distribution, but there is a great increase in movement and in the shifting of home ranges amongst mature shrews, particularly males. This helps to bring the sexes together, which is important bearing in mind that females are only receptive during the brief period of oestrus, which lasts only 2–4 hours. It must be a selective advantage for males to contact as many females as possible, and as frequently as possible, to be sure of matings. It is interesting to note that studies suggest that no mature female fails to be mated, and that there is a high degree of fertility amongst both males and females. So their strategy for mate location must be a successful one, despite their asocial nature. The nomadic behaviour of males may have another effect—promoting genetic mixing amongst and between populations of shrews.

The increase in the home ranges of breeding females exemplified by *Sorex vagrans* and *S. obscurus* may be the result of the increased energy requirements of pregnant and lactating shrews, and a need for larger foraging areas for themselves and their developing offspring. However, the more fluid boundaries of these females' home ranges suggests that they are too busy rearing their young to maintain and patrol a regular territory at this time.

Some species are more sociable than those discussed so far, and may tolerate close association with their fellows. The white-toothed shrew *Crocidura russula*, for example, has quite small home ranges (75–395m^2)

which overlap considerably with those of neighbouring shrews. It is doubtful that these shrews are territorial at all. Some are even positively gregarious; *Cryptotis parva* for example, which will live in small colonies with several adults sharing a nest, a latrine and a home range. In captivity, as many as twelve individuals have been found sharing a nest, with no sign of aggression. Even breeding females will rear their young while sharing a cage.

POPULATION DENSITIES AND HOME RANGE SIZE

Population densities of shrews are extremely variable, and depend on the species, the time of year and the nature of the habitat. Populations of *Sorex arcticus*, for example, have been correlated with the depth of the water table below ground: the wetter the site, the greater the problems of flooding and the lower the population size of shrews (Buckner, 1966a).

Population density may also vary with differences in food availability and vegetation cover, although these are often difficult to quantify. The variable nature of shrew population sizes is exemplified by the common shrew (*Sorex araneus*), which has been the subject of many field studies in a number of different habitats. The results of some of these studies can be seen in Table 3.4.

Table 3.4 Estimates of population densities of *Sorex araneus* in different habitats (numbers per ha)

Summer	Winter	Habitat	Author
49–62	–	Deciduous woodland, England	Crowcroft, 1954
69	17–27	Deciduous woodland, England	Shillito, 1960
18	12–13	Dune scrub, Netherlands	Michielsen, 1966
8	5	Grassland, England	Pernetta, 1977a
7–21	5–9	Scrub-grassland, England	Churchfield, 1979
41	6	Grass banks, scrub, England	Churchfield, 1984a
43–98	53–65	Grassland, England	Churchfield and Brown, 1987
6–52	–	Various habitats	Aulak, 1967

As we have seen, shrew populations are typically high in summer when breeding occurs, and lower in winter. They also vary considerably from year to year (see Figure 2.5), although they rarely seem to suffer the regular peaks and crashes that are evident in populations of small rodents.

The presence of other shrew species may have an influence, too. Where *Sorex arcticus* and *S. cinereus* occur together, their populations have been

Table 3.5   Examples of population densities estimated for different species of shrew

Species	Density per ha	Habitat	Author
Britain & Europe			
Sorex araneus	17–69	Deciduous woodland, England	Shillito, 1960
S. araneus	12–18	Dune scrub, Netherlands	Michielsen, 1966
S. minutus	25–40	Dune scrub, Netherlands	Michielsen, 1966
S. minutus	5–30	Grassland, England	Churchfield and Brown, 1987
S. minutus	25–40	Grassland, Ireland	Ellenbroek, 1980
Neomys fodiens	<3	Grass/shrub/watercress beds, England	Churchfield, 1984a
Crocidura russula	77–100	Around villages, Switzerland	Genoud, 1978
N. America			
Sorex vagrans	12	Mixed forest, British Columbia	Hawes, 1977
S. obscurus	12	Mixed forest, British Columbia	Hawes, 1977
S. cinereus	1–23	Tamarack bog, Manitoba	Buckner, 1966a
S. arcticus	0–10	Tamarack bog, Manitoba	Buckner, 1966a
Blarina brevicauda	0–5	Tamarack bog, Manitoba	Buckner, 1966a
B. brevicauda	3–30	Various habitats	Novak and Paradiso, 1983
Cryptotis parva	31	Forest, Florida	Kale, 1972
Notiosorex crawfordi	24	Desert scrub, Arizona	Hoffmeister and Goodpaster, 1962

found to vary inversely with each other (Buckner, 1966a). Some indication of the population densities of different species is given in Table 3.5.

As population densities of shrews are extremely variable so, too, are the sizes of their home ranges. Table 3.6 shows some examples of home ranges for different European and North American shrews which have been the subject of detailed field studies. Some species, such as *Cryptotis parva* and *Blarina brevicauda*, seem to maintain very large home ranges, occupying areas of up to 12,000m^2 and 8,000m^2 respectively. Others, such as *Sorex araneus*, have quite compact home ranges, occupying as little as 400m^2. However, even in these species the diameter or length of the home range may vary between 20m and 75m in the same habitat. There is generally no difference between the sexes, except during the rut when males expand their home ranges enormously. Home range size is not directly correlated with body size. One of the smallest species, *S. minutus*, always has a home range 1.5–2.2 times larger than

Table 3.6   Examples of home range size (in m²) in different species of shrew

Species	Winter	Summer	General	Author
Sorex minutus	900–1,850	530–800		Michielsen, 1966
S. minutus	1,400–1,700	–		Pernetta, 1977a
S. araneus	500–600	400–450		Michielsen, 1966
S. araneus	2,800	–		Buckner, 1969a
S. araneus	800–1,100	–		Pernetta, 1977a
Neomys fodiens	77–173	101–373		Lardet, 1988
Crocidura russula			75–395	Genoud, 1978
S. vagrans	510–1,986	732–5,261		Hawes, 1977
S. obscurus	338–1951	605–4,425		Hawes, 1977
S. cinereus			5,549	Buckner, 1966a
S. arcticus			5,913	Buckner, 1966a
Blarina brevicauda			3,929	Buckner, 1966a
B. brevicauda			2,000–8,000	Banfield, 1974
Cryptotis parva			4,400–12,000	Choate and Fleharty, 1973

S. araneus, with which it coexists, although it is some three times smaller than the latter. The explanation for this may lie in the fact that *S. araneus* maintains more of a three-dimensional home range which includes subterranean tunnels as well as the ground surface. *S. minutus*, on the other hand, is primarily a ground-surface dweller, and may need to cover a larger area in the search for prey.

Home ranges differ in size according to season and age, as we have seen. For instance, juvenile common shrews in summer have a mean home range length of 37m but for sub-adults in winter it is 29m. Apart from seasonal and age considerations, the two most important factors which affect home range size are population density and the nature of the habitat, particularly regarding food supply. If population density is high then there is a tendency for home ranges to be smaller, and the amount of overlap may change. For example, *S. cinereus* shows great intolerance of infringements of its home range at all population densities. In order to accommodate increasing numbers of shrews and survive competition from conspecifics, individual home ranges become squeezed and reduced in size at high population densities. *S. arcticus*, on the other hand, is prepared to tolerate an increasing amount of overlap as population density arises, perhaps in order that individuals can maintain a reasonable home range size (Buckner, 1966a). Differences between these species in the overlap tolerated at different population densities are demonstrated in Figure 3.4.

There is also some evidence that home range size is increased as food availability decreases. It is noticeable that North American species generally have larger home ranges than the common European and British species: compare *S. araneus* and *S. minutus* with the North American *S. vagrans* and *S. obscurus* in Table 3.6. This may be related to resource availability. Many of the studies of the spatial distribution of North American shrews have been carried out in relatively resource-poor habitats such as forests and tamarack bogs which also suffer marked seasonal changes in prey numbers, with a great reduction in winter. Population studies of European species, on the other hand, have often

Figure 3.4 Home range overlap tolerated by Sorex cinereus *and* S. arcticus *at different population densities (after Buckner, 1966a)*

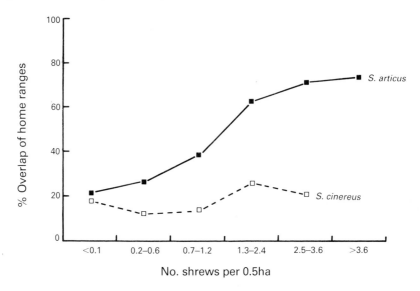

been conducted in relatively resource-rich grasslands and scrublands where prey numbers can support much higher populations of shrew. However, quantitative evidence of the effects of prey availability on home range or territory size is hard to come by.

Shrews do seem to react to changing availability of prey. Holling (1959) found that an increase in the density of a *Sorex cinereus* population coincided with an increase in prey densities. In his study of short-tailed shrews contained in large enclosed areas of grassland, Platt (1976) found that a decline in the population of the field vole (*Microtus*), was also accompanied by a reduction in shrew density, together with an increase in the home range size of these shrews. Voles are an important prey of short-tailed shrews, especially in winter. At high prey density, the area occupied by resident shrews was only 300–700m^2, but at low prey density this area increased to 1,000–2,200m^2. As prey density decreased, shrews also frequently shifted their home ranges, moving from areas of low prey density to better areas. Outside the enclosed areas, where prey density had not declined, the numbers of resident shrews remained quite constant and home ranges did not increase or shift position. As prey availability increased, the size of the home ranges decreased, both inside and outside the enclosures.

With high population density of shrews and abundant invertebrate prey, home ranges overlap considerably. During some of my recent studies of common shrews in invertebrate-rich grassland in southern England, population densities rocketed to 200 per hectare and never fell below 25 per hectare. Home ranges were concentrated in certain areas of the grassland, even though there was room to expand into neighbouring parts, and they overlapped greatly. There was no sign of territoriality, even in winter.

Shrews have home ranges of all shapes. In terrestrial species they may be roughly oval, triangular or squarish, depending upon the influence of their neighbours or on the vegetational and topographical features which may act as barriers or convenient borderlines. Shrews inhabiting hedge-rows have long, linear home ranges, venturing little outside the shelter provided there.

Aquatic species such as *Neomys fodiens* and *Sorex palustris* also have linear home ranges based on the banks of streams and rivers. They occupy a length of river-bank plus the adjacent area of water and, often, a small area of hinterland. *N. fodiens* typically has a home range length of 20–93m alongside a stream.

INTRASPECIFIC BEHAVIOUR OF SHREWS AND MAINTENANCE OF TERRITORIES

A greater insight into the actual behaviour of shrews in their home ranges and territories, particularly with respect to the presence of intruders, came from a study of the social organisation of the American short-tailed shrew (*Blarina brevicauda*) by Platt (1976). He adopted something of a compromise in his study by investigating wild shrews in large outdoor enclosures or pens containing soil and vegetation. Some of the enclosures were 4,050m^2 in area, large enough to support a shrew in a home range similar in size to that in the wild. Between the wall of the enclosure and the grassy interior, a bare area 0.6m wide was maintained. The movements of shrews were tracked by means of radioactive tagging. Each shrew was given a small wire tag containing the radioisotope tantalum 182. In this way Platt could monitor the reactions of resident shrews to intruders and vice versa, and compare the behaviour of shrews introduced to occupied and unoccupied enclosures.

After release into an enclosure containing a resident shrew, an intro-duced animal would move to the periphery of the vegetated area and remain there. It moved quickly along the border, making only short sorties into the interior of the pen, and occasionally out into the bare area at the edge. These intruder shrews remained very active for several hours after their introduction, until they had their first encounter with the resident. Shrews introduced to an unoccupied pen, on the other hand, were active for only an hour or so, and quickly settled down.

Resident shrews moved freely through the whole area of the enclosure. They detected the presence of an intruder as they moved into an area previously passed through by the stranger, presumably by scent. As soon as detection of the stranger had occurred, the resident became much more active, and interactions between resident and intruder occurred fre-quently, although they lasted for only ten minutes or less on each occasion. An interaction would commence with screaming by the resi-dent, starting some 30–100cm from the intruder, followed by attacks and chases. As the intruder was attacked it would often roll on to its side or back, and the resident would bite at its belly. They often became locked in a ball, rolling over the ground as they kicked and fought. Eventually the intruder would flee and the resident give chase, nipping at its sides or rump. Hostilities would suddenly cease as the intruder retreated to the bare area at the edge of the enclosure. The resident usually remained amongst the vegetation but would run up and down in the vicinity of the intruder. The intruder frequently attempted to re-enter the vegetated area, only to be rebuffed by vocalisations from the resident, quickly reinforced by attacks and chases if the intruder persisted.

Attacks could result in serious injury, even death, to the intruder, and so they had to be watched carefully and removed when necessary. However, some introduced shrews did establish themselves in an already occupied pen. In one trial involving two males, there were numerous interactions at first, but the introduced male adopted a dominant stance and himself carried out attacks on the resident. They engaged in vocal sparring but attacks declined. The two shrews then partitioned the enclosure between them, and no further interactions ensued. A similar result was achieved when a male was introduced to a pen occupied by a female. A trial involving two females also resulted in partitioning of the pen, and a male which was subsequently introduced was able to move over the whole area.

In an area of grassland outside the enclosures, where shrews could be tracked under more natural conditions, Platt found that the home range boundaries of adjacent residents of either sex were contiguous, but shrews occasionally ventured into each others' areas. If one shrew emigrated, its neighbour began moving along the common border the next day and briefly entered the vacated home range. After 48 hours it began to forage inside the area. Residents seemed reluctant to enter occupied home ranges but were quick to exploit unoccupied areas, as was indicated in studies of *Sorex araneus* and *S. minutus*. Nomadic shrews behaved much like the intruder shrews in the enclosure experiments: they were highly active and reluctant to move into occupied areas.

Residents showed considerable activity at the borders of their home ranges, and a search of the enclosures following the trials revealed that there were more faecal deposits at the edges of the home ranges compared with the rest of the area. Nomadic shrews, on the other hand, did not mark the boundaries of recently invaded areas. They tended to move along the edges of occupied areas and avoid interactions with residents in the centre of the home range. The importance of faecal marking was further explored by observing the behaviour of captive shrews at closer quarters. A cage was subdivided with a wire mesh partition, and one half of the cage was occupied by a resident short-tailed shrew. Within a few hours of a new shrew being introduced into the unoccupied half of the cage, the resident began to deposit faeces along the partition. Within 24 hours the newly introduced shrew also began to deposit faeces on its side of the partition. Both shrews spent much time at the partition during the first day, sniffing and licking the faeces, but after that little time was expended there, and there was no confrontation between the shrews. It seems, then, that faeces do provide information about occupied areas which is recognised, and respected, by conspecifics. More will be said shortly about the importance of scent in communication between shrews.

Common shrews (*Sorex araneus*) are well known for their pugnacity, and captive animals introduced into a cage already occupied by another individual will be attacked vigorously by the resident. Scuffles and chases will persist whenever the two meet, and may result in the death of one, usually the non-resident. Such pugnacity even extends to the reaction of a resident shrew to the introduction of the stuffed skin of a dead shrew. The resident in one such test soon detected the presence of the 'intruder' and attacked it repeatedly, biting it viciously until it had been torn into pieces. However, this overtly aggressive behaviour occurs under conditions of great stress in captivity, where there is no escape for either shrew, a situation very different from that in the wild. But fights do certainly occur amongst wild shrews. I once witnessed two common shrews kicking, biting and squeaking as they rolled around together on a hedge bank. So involved were they that I was able to pick one up by the tail while they continued to fight as I held them in mid-air, locked together in combat.

Other species may learn to live in close proximity to each other. When water shrews (*Neomys fodiens*) are kept together in a large cage, the initial scuffles are gradually replaced by long, low vocal churls acting as a warning for each to keep its distance. They adopt different areas of the cage and separate nests, although there may be periodic incursions into each others' nest sites followed by loud shouting matches and the eventual retreat of one or the other.

It seems that shrews of different ages react differently to each other, and that some form of hierarchy or dominance occurs. Laboratory studies suggest that young common shrews act aggressively towards old males and that the latter are more likely to retreat from an aggressive encounter than the juveniles. These old males may then be forced to move out and occupy poorer habitats. Breeding females, on the other hand, seem to be socially dominant to both the adult males and the juveniles (Moraleva, 1989). Adult males, once they assume a nomadic existence in the breeding season, rapidly have their old home ranges snapped up by juveniles and become socially inferior to them. It is speculated that the social pressure of increasing population density and the accompanying aggressive behaviour of young shrews may be a prime cause of the eventual death of the old males (and probably the old females too, once they have bred).

INTERSPECIFIC BEHAVIOUR

How do different species of shrew coexisting in the same area react to each other? Common and pygmy shrews often have overlapping home ranges and territories and they seem to avoid each other. When they do meet, *Sorex minutus* is submissive towards *S. araneus* and rapidly and quietly retreats before a confrontation. If a captive pygmy shrew is introduced to a cage already occupied by a common shrew, it may be chased and attacked on contact.

Mutual avoidance seems to be practised between other species when they meet. When individuals of the African species *Crocidura hirta* and *Myosorex varius* were introduced into a cage with *C. flavescens*, all three avoided each other. *Sorex palustris* and *S. vagrans* placed together in the same cage showed some agitation at first, and when they met they squeaked and occasionally fought momentarily. But both seemed keener to avoid each other than to engage in a fight, the small *S. vagrans* showing most effort to avoid the larger water shrew.

Such a system extends to encounters between shrews and other small mammals. When *S. palustris* and the deer mouse, *Peromyscus maniculatus* were kept together in the same cage, they used the same food dishes and the same tunnels, but tended to avoid each other. Occasionally, the deer mouse would attack the shrew if they encountered each other in the more open areas of the cage. Mostly when they met, the shrew let out loud squeaks and then one of them would quickly retreat or wait quietly until the other had gone away.

An alternative mechanism of minimising contact is interspecific territoriality. Hawes (1977) found evidence of interspecific territoriality leading to competitive exclusion which helped to keep *Sorex vagrans* and *S. obscurus* separate in areas where they coexisted. She noticed that the exact boundaries of the home ranges of neighbouring shrews of different species followed more closely the edges of an adjacent home range of the opposite species than those of any topographical or vegetational changes.

The problem of different species encountering each other in their daily lives may be further avoided by each adopting a slightly different micro-habitat. For example, *S. obscurus* was captured predominantly in areas characterised by mor soil (acid, with a distinct humus layer) and its accompanying vegetation, while *S. vagrans* generally occurred in areas of moder soil (less acid, with a greater mixing of organic matter with the mineral soil). Year after year, despite the turnover in population and the birth of new individuals, each species settled primarily in the same distinct areas of Hawes's study site, which coincided with these different soil and vegetation types. Those shrews which settled in the 'wrong' area often disappeared before the breeding season. Such subtle differences in micro-habitat selection have been found for a number of other shrews, for example the closely related and very similar *Sorex araneus* and *S. coronatus* in areas of overlap in Switzerland (Neet and Hausser, 1990). More will be said about habitat selection in shrews in Chapter 7.

FUNCTIONS OF TERRITORIES IN SHREWS

It has been speculated that territoriality acts as a regulatory mechanism preventing excessive population densities. Michielsen (1966) found evidence of high population pressure amongst shrews, forcing some individuals to settle in sub-optimal habitats which led to high mortality of old adults and young shrews in late summer and autumn. Hawes (1977) deduced that success in overwintering and reaching sexual maturity was highly correlated with the ability to establish a territory in autumn and hold it through the winter. She suggested that winter food levels served as a delayed density-dependent factor regulating populations of shrews. This may be so in resource-limited habitats in winter, such as the forest inhabited by *S. vagrans* and *S. obscurus* on which Hawes worked in Canada. Here, shrew densities were very low, only some 12 per hectare, suggesting that conditions were not very favourable for them.

That territorial behaviour under such circumstances conveys a selective advantage to shrews is supported by a study of the ecological energetics of the short-tailed shrew, *Blarina brevicauda* (Randolph, 1973). This is a relatively large shrew whose winter food requirement was estimated to exceed the summer requirement by some 43 per cent. While only about 3 per cent of the total prey resource was exploited by this shrew population in summer, this increased to 37 per cent in winter. But food availability on the forest floor was some eight times less in winter than in summer. So declining food availability coupled with increased needs in winter may encourage territoriality in these shrews.

Territoriality provides a mechanism for sharing resources and minimising contact between individuals. This may be particularly important in habitats where food resources are limited and suffer seasonal shortages. Shrews are unable to hibernate so must remain active all year round. The ability to occupy a territory and deter intruders may therefore be an important survival strategy. Since shrews are the major predators of invertebrates on the ground in many terrestrial habitats, defending an area against other shrews would certainly make more food available to the territory owner. However, the advantages of holding a territory must be balanced against the costs of defending it and chasing away invaders. If food resources are very scarce, a large territory is needed, which demands much time and effort to patrol, using up valuable energy. Similarly, if population density is very high and there is constant invasion of home ranges by neighbours and nomads, too much effort may be required to maintain a territory. When food is plentiful there may be enough for all to share without developing a system of territoriality. So to make the best use of their energy and the resources available to them, shrews must be flexible in their system of territoriality. There is evidence that they are territorial when it is advantageous to be so, such as in winter, but not otherwise. In resource-rich habitats, territoriality seems to break down.

Shrews wander widely over their home ranges in their daily foraging activities, and so it is probable that territory patrol and assessment of the neighbours' activities can be combined with foraging, at little additional cost. They leave scented faeces and urine together with odoriferous substances from specialised scent glands as they pass along which must help to reinforce ownership of the area, as well as give information to and about invaders. Home range residents are also dominant over intruders, and highly aggressive. All this probably helps to keep individuals apart and confrontations to a minimum.

4 Communication and orientation

Shrews employ their full range of senses in communication and orientation, including smell, hearing, sight and touch, but the relative importance of each depends very much on the circumstances. By far the most important means of communication between shrews are scent and vocalisations.

SCENT

Like other mammals, shrews have scent glands scattered in the skin over most of the body. The normal sebaceous glands associated with each hair follicle are important in lubricating, waterproofing and protecting the skin and the hair. Sweat or sudoriferous glands also occur but vary in size and distribution. They are concentrated in the skin of the ventral surface of the shrew, particularly the thoracic and abdominal areas but, compared with sebaceous glands, they are small and relatively few in number. Few if any sweat glands occur on the dorsal surface of shrews. The sudoriferous and sebaceous glands in mammals, including shrews, are often concentrated in certain areas of the body and modified to perform an additional function as scent glands by producing specific and distinctive odours. These glands have an important role in communication between individuals of a species, particularly serving to indicate reproductive condition or to help mark ownership of territories and home ranges. The scent glands of shrews probably have similar functions, but their precise role is not fully understood. A summary of the scent glands, together with the shrews found to possess them, is shown in Table 4.1.

The Flank Glands

Shrews possess a number of different scent glands on their bodies but the most obvious in all species are the paired lateral or flank glands. These are small oval areas of skin approximately midway between the fore and hind legs, just posterior to the ribs, one on each flank. They are well supplied with blood vessels and, when viewed from the inner skin side, they appear as reddish, slightly thickened areas. They are commonly

Table 4.1 Examples of the scent glands possessed by shrews

Type of gland	Location	Occurrence	Species
Flank glands	One on each side between fore and hind legs	Present in both sexes, but rudimentary in females; prominent and active in mature males	*Sorex* spp
Flank glands	One on each side between fore and hind legs	Active in all males; less prominent or rudimentary in females	*Blarina brevicauda*
Flank glands	One on each side between fore and hind legs	Active in males and females of all ages	*Suncus murinus*
Ventral gland	Centre of abdomen	Both sexes; most developed in males	*Blarina brevicauda* only
Chin, throat, neck and belly glands		Both sexes	*Crocidura* spp
Post-auricular gland	Behind ears	Both sexes	*Crocidura* spp *Suncus murinus*
Caudal glands	Base of tail	Lateral caudal gland in both sexes; sub-caudal gland only prominent in mature males	Some *Crocidura* spp Not in *Sorex*
Anal glands	Anal region	Possibly in all shrews, scent-marking faeces, occurrence not confirmed	Probably in all species

about 4mm wide by 8mm long in the smaller *Sorex* species such as *S. araneus* and *S. vagrans*, but they vary somewhat in size from species to species, and in the American short-tailed shrew (*Blarina brevicauda*) they are up to 13mm long. They are typically bordered by short, stiff hairs and, when fully developed, they produce a highly odoriferous, greasy secretion which sticks to these hairs, creating a small, dark, moist patch.

Close examination of sections of these flank glands under the microscope reveals that they are composed of a mass of large sweat glands lying in the deeper part of the dermis of the skin and, above them, of large sebaceous glands in the upper dermis. The sebaceous glands differ little from the normal sebaceous glands found elsewhere in the skin, although they may be larger. The sweat glands are mostly larger than normal ones, and are massed closely together. The fine structure of the flank glands differs somewhat from species to species. In many *Sorex* species, such as *S. vagrans*, *S. fumeus* and *S. trowbridgii*, these glands are composed of enlarged sweat glands and sebaceous glands in approximately equal proportions. In *Suncus murinus*, the large Asian musk shrew, they consist mostly of sebaceous glands and relatively few sweat glands. In *Sorex vagrans* and *S. fumeus*, the hair follicles over the scent gland slant towards the centre so that the hairs are opposed to each other over the surface of the gland. *S. fumeus* has particularly prominent flank glands which are clearly delimited by a circular ridge of raised skin caused by the local accumulation of sweat glands and surrounded by a small pit containing a dense growth of short hairs.

It seems that it is the sweat gland elements which produce the distinctive odour while the oily secretions of the sebaceous tubules function to trap the odour and make it more persistent.

These flank glands are conspicuous even in very young shrews. Although they are not functional at this age, they are clearly visible in infants as young as five days, before they develop a full covering of fur. However, they show marked differences in their degree of development and activity with the age, sex and breeding condition of the shrew, and the same pattern is not seen in all species. In many species of shrew, the flank glands, although present in both sexes, seem only to reach maximum development and secretory activity in mature, breeding males.

Hawes (1976) conducted a study of the flank glands of *Sorex vagrans* and *S. obscurus* in North America and made a number of important observations. She found that the glands of female shrews of all ages, and of immature shrews of both sexes, were merely rudimentary. In the immatures, the glands were not even visible externally. In breeding females the glands could be detected as small oval areas only on the inner side of the skin, and she noted no change in the odour of these shrews as the breeding season advanced. The glands were, however, highly developed in breeding males, coincident with their growth to sexual maturity. In January and February, when the testes began to enlarge, the hair at the site of these glands moulted and the glands themselves began to develop and became evident as a moist yellowish-brown patch on each flank, about 8mm long, covered in short bristles and exuding a strong, distinctive odour. Their smell increased with excitement, and placing two shrews together in a cage resulted in a wave of odour emission. Not only that, but Hawes's own nose could detect a difference in the odours emitted by the two species: *S. vagrans* exuded a musky odour while *S. obscurus* emitted a characteristically pungent, acrid secretion. So distinctive was the difference that she could easily distinguish the two species by their smell without even looking at them.

A similar pattern of development of the flank glands is seen in a number of other species, including *Sorex araneus* and *S. minutus*. Here, these glands are just visible externally in immature shrews of both sexes, appearing as small oval patches of short, bristly hairs, often slightly darker in colour than the surrounding pelage, but there is little if any sign of secretion. As the animals grow towards sexual maturity the glands become more easily visible but only in the mature males does the hair become oily and matted at the site of the gland with a very distinctive and pungent, musky odour. The odour is particularly noticeable when these adult males are excited or handled. It is hardly noticeable at all in younger shrews and mature females. In *Sorex fumeus* and *S. murinus*, however, these glands are prominent in both sexes throughout the breeding season.

In the American short-tailed shrew, *Blarina brevicauda*, the flank glands show a slightly different pattern of development. Eadie (1938) reported that there was some variation in the state of the glands with the reproductive cycle, at least in males, as with *Sorex* species. They appeared to increase in size and activity in males concurrent with the growth of the testes as the breeding season approached, while in females there was no change. However, on closer scrutiny, Pearson (1946) found that the glands of breeding males were, in fact, scarcely larger than in non-breeding males although they might appear more developed because of an increase in vascularity (blood supply) at the site of the glands. He made sections of the glands of shrews of different ages and sexes which revealed that they were most highly developed and functional in both breeding and immature males, and rather less so in immature females. They were very poorly developed in oestrous, pregnant and lactating females, with pregnant animals having much reduced glands. He went on to investigate the effects of sterilisation and hormone addition on the growth of the glands. Sterilisation did not adversely affect their development. Instead, sterilised males and females had highly developed flank glands. Injections of testosterone increased the vascularity of the glands in both sexes, but there was no increase in the size of the sebaceous or sweat gland elements. Neither did injections of oestrogen cause a regression of the glands in either sex.

It has often been suggested that the distinctive odour of shrews makes them unpalatable to predators, and that the flank glands could have an important role in this respect. However, shrews are frequently killed and occasionally eaten by a range of predatory mammals such as foxes, weasels and bobcats, and they are regular prey items of raptorous birds such as owls and hawks. It is reported that some 23 bird species are known to prey upon *Sorex* species and 27 species on *Blarina*. Their smell may repel mammalian carnivores, but it certainly does not deter birds of prey, the main predators of shrews, which in any case have a relatively poorly developed sense of smell. Moreover, if the flank glands were intended to function as an anti-predator device it is strange that pregnant and nursing females, the most valuable and vulnerable section of the shrew population, should have the weakest scent glands, even non-functional in certain species.

It seems most likely that the flank glands function in the social life of shrews. The strong correlation between the growth of the testes and the development of flank glands in males, at least in species of *Sorex*, suggests that there is a reproductive role for the odour produced. Shrews have poor eyesight and, in the cluttered environment in which they live, it might be difficult for them to locate each other and, particularly, to find mates in a suitable reproductive condition. So perhaps scent assists in this, acting as a pheromone over quite large distances and advertising the sexual condition of individuals. Yet this could only function effectively in males, since the females of most species have poorly developed flank glands and are much less smelly than males. If the odour was to function as a sexual attractant, then it would seem more advantageous for females to produce a characteristic scent for the brief period when they were in oestrous and receptive to males. As it is, females of all species are least smelly at the very time that it would be of greatest advantage. Perhaps, then, the distinctive scent of the males advertises their presence and provides a sexual attractant to females, making them less aggressive and permitting the approach of males into the females' home ranges or territories. At the same time the heavy scent may deter other males, staking a claim to a number of females' territories.

It seems most likely that the scent glands have a dual role. Being generally smelly creatures, it is probable that shrews leave their scent wherever they go, brushing against the vegetation and defecating in strategic places, so informing other shrews not only of their presence but of their reproductive condition. In *Suncus murinus*, the Asian musk shrew, both sexes have active flank glands and captive animals have been observed to mark the sides of their cages and also objects in their cages by secretions from the flank glands, suggesting that they function primarily in scent-marking and may have a role in territoriality. So, scent probably functions in keeping individuals apart by advertising their presence and ownership of pathways, tunnels and nests, and hence in the establishment and maintenance of home ranges and territories. However, it is obvious that the flank glands do not only serve in this capacity since they are least active in immatures of most species during the autumn and winter, at the very time when shrews exhibit territoriality and when the scent would be most useful in deterring others and reducing competition for space and food.

That different species produce different, often distinctive, odours suggests an additional role, one of species recognition. Hawes (1976) noted the different odours of *Sorex vagrans* and *S. obscurus*, and Skaren (1964) also found differences between *S. araneus* and *S. unguiculatus*. *Neomys fodiens*, too, has a very distinctive smell. Perhaps all shrews have species-specific smells which evolved as a useful and necessary recognition signal, of particular importance in areas where several similar species are sympatric, helping to keep them separated and avoiding competition. Individuals would also not waste time soliciting mates which may turn out to be of the wrong species. But, again, any conclusions on the involvement of the flank glands are very tentative because of the relative absence of odour from them in adult females, and also in immatures in autumn and winter when competition should be greatest.

The Ventral Gland

While all shrews possess lateral flank glands, only one species has been found to have a ventral gland, namely *Blarina brevicauda*, the American short-tailed shrew. This gland comprises an oval area of skin about 10mm wide by 30mm long in the centre of the abdomen, just anterior to the inguinal region. It is readily visible in all these shrews because the hairs covering it are quite different from the rest of the pelage. Instead of the long, dark body hairs, those covering the gland are quite short, fine, wavy and pale in colour. As with the flank glands, the area is frequently moist and greasy. The skin here is thickened and well supplied with blood vessels, features which are visible from the inner surface of the skin but not as obviously as with the flank glands. The ventral gland rests on a layer of connective tissue and muscle and the skin actually adheres more tighly to the underlying muscles here than elsewhere on the body.

The ventral gland consists primarily of enlarged sebaceous glands and their secretion is formed, at least in part, by the breakdown of the cells nearest the centre of the gland. The sebaceous glands open into the hair follicles and there are usually two or more glands per follicle, so there is a great concentration of them in this area. Sweat glands, on the other hand, are fewer in number, although they are enlarged here compared with other areas of the skin. They open directly on to the skin surface rather than into a hair follicle. So, it seems to be the sebaceous elements which are the most important in the structure and functioning of the ventral gland while the sudoriferous elements are more important in the flank glands.

While this ventral gland is not found in other shrews, a structure rather similar to it is found in several mustelids, which are notoriously smelly animals.

As with the flank glands, there is some variation in the size and secretion of the ventral gland throughout the year and according to the sex and reproductive status of the shrew. Both sexes possess this gland. In males, it actively secretes all year round, although it has been reported that the sebaceous elements grow slightly larger as the testes increase in size towards the breeding season (Eadie, 1938). However, Pearson (1946) found that the gland was as well developed in terms of the overall thickness and size of the sebaceous and sweat glands in males with small testes as in males in breeding condition. The ventral gland of females is always slightly thinner and smaller than in males, and in pregnant animals it shows a marked decrease in function: the sebaceous elements are small and the skin over the whole glandular area reduced in thickness and there is little secretory activity. Otherwise, the gland seems to function all through the year in both sexes, but more prominently in males.

The function of the ventral gland may be different from that of the lateral glands, since it has a slightly different structure and level of activity in both sexes. Again, it has been suggested that it has a protective function to repel predators. The secretion from the gland can easily be obtained by pressing a finger on the area; an oily and highly odoriferous substance is produced. Unlike the flank glands, the ventral gland is strongly and closely attached to the underlying abdominal muscles. Elsewhere on the body the skin is quite loose. It is possible, then, that strong contraction of the abdominal muscles would exert pressure on the components of the gland, causing the secretion to be extruded. So, the shrew might be able to produce a waft of repellent odour voluntarily when frightened or angry. However, this theory does not adequately take into account the fact that the gland functions more prominently in males and shows a marked decrease in activity in pregnant females, which would have the most to gain by such a defence mechanism. So, this gland, too, probably has some function in the social life of short-tailed shrews which has yet to be fully understood.

Other Scent Glands

Shrews generally give off a strong odour, but this may not be due simply to the flank glands, since they are usually only actively secreting for a limited period as shrews mature. White-toothed shrews (*Crocidura* spp) give off a sweet, musky odour at all times. If the skin containing the flank glands is removed from males of *Suncus murinus*, they continue to produce the same, strong distinctive odour. The odour seems to issue from concentrations of sweat glands on the throat (throat gland) and behind the ears (post-auricular glands). Scent from these glands can be produced suddenly and in such high concentrations that it can be detected several metres away. Their activity is dependent upon sex hormones in both males and females. Castration promptly results in atrophy of the glands, and no further odour production. Treatment with testosterone or oestradiol induces the recovery of the glands and resumption of odour production in both castrated males and females.

Females of the least shrew (*Cryptotis parva*) also possess well-developed glands anterior to the ventral part of the aural cartilage which secrete most actively during pregnancy. It is not known whether males possess this gland too.

White-toothed shrews such as *Crocidura russula* seem to possess small scent glands on the belly, chin and neck, since they have been observed to scent-mark using both these areas of the body. *C. russula* also possesses subcaudal and lateral caudal glands at the base of the tail. The subcaudal gland, on the ventral surface of the tail, is particularly well developed, causing a thickening of the tail there. It comprises enlarged sebaceous and sweat glands resembling the structure of the flank glands. It is only prominent in males as they attain sexual maturity, and so appears to have a sexual function as do the flank glands in mature males of many species. The gland has not been found in other closely related *Crocidura* species such as *C. suaveolens* and *C. leucodon*. Neither has it been found in *Sorex* species, although there is evidence of some secretory activity at the base of the tail suggesting the presence of some glandular structure.

As we have seen, strategic positioning of the faeces may be important in the marking of home ranges and territories, particularly at the boundaries suggesting that they are impregnated with scent peculiar to that individual shrew. Certainly the faeces possess the same characteristic smell as the shrew itself. Scent glands associated with the anal area (circumanal and proctodaeal glands) have been found in males and females of species of *Sorex*, *Crocidura*, *Suncus* and *Neomys*, and are probably possessed by all shrews.

A white secretion is often produced from the eyelids of *Sorex* and *Crocidura* species when they are agitated or excited, implying the presence of a gland associated with the eye, but the function of this is unknown.

In addition to the range of scent glands described, odour is probably produced from the general body surface (especially the ventral region) via specialised cells distributed over the surface of the skin. This may help to provide a shrew with its own personal identity when other scent glands are not actively secreting.

Scent-Marking Behaviour

Despite possessing such an array of scent glands, specific scent-marking behaviour has rarely been observed in shrews and so the precise manner in which scent is employed by these animals can often merely be speculated upon. The way in which the flank glands are used by *Sorex* species, for example, has not been clearly defined. It is assumed that scent from these organs is deposited passively as animals pass through and brush against the vegetation and the sides of their burrows.

However, specific scent-marking behaviour by different crocidurine shrews has been observed in more detail. Three methods of marking were observed in the African species, *Crocidura flavescens*, in captivity (Baxter and Meester, 1982). Both sexes displayed marking behaviour but males did so much more frequently. The most common method was by using the flank glands, usually by leaning against an object and moving the body backwards and forwards. Another way of marking involved rubbing the chin and neck against the substrate of an object, and this always occurred after an aggressive encounter with another shrew. More rarely, marking using the anal area was observed, again after an aggressive encounter: the anal area was lowered against the ground and the body dragged forward by the fore limbs. Occasionally the whole ventral surface would be placed against the ground and the body wriggled forwards so that all the scent-producing areas could be utilised at the same time.

Other *Crocidura* species have been observed to indulge in similar scent-marking behaviour. For example, captive *C. suaveolens* have been seen to 'belly mark' both in their own familiar areas and in unfamiliar places: the belly is pressed to the ground and dragged forward with the fore feet. 'Chinning' has also been observed, the underside of the chin being rubbed against prominent objects in much the same way as rabbits do.

Marking behaviour may be dependent on the reproductive state of the shrew. For example, males of *Suncus murinus* showed a decrease in marking with the throat, flank and anal glands within three or four weeks of castration. Subsequent treatment with the hormones androgen and oestradiol completely restored their scent-marking behaviour.

Captive shrews tend to deposit their droppings in faecal middens in the far corners of their cages or enclosures. Sometimes they place them around the edges of cages, and on the walls and on the outsides of their nests, and some species, including *Sorex palustris*, will quite purposefully deposit faeces on food items including cached prey. Specific patterns of defecation have been observed in captive white-toothed shrews. They may stand on their fore paws with the hindquarters pressed against the wall of the cage and deposit faeces on the wall, several centimetres above ground level. While urinating, *Suncus murinus* may drag itself forward with its hindquarters against the ground, and urine marks can be seen as trails on the cage floor.

Strategic placing of smelly faeces around the borders of a territory or home range, and in other prominent places within it, is typical of many territorial mammals. As we have seen in the studies of territorial behaviour in shrews such as *Blarina brevicauda*, there is evidence that faeces are used to proclaim ownership of an area and delineate the boundaries.

Functions of Scent in Shrews: Summary

There is still much to be understood about the functions of scent in shrews. Undoubtedly, odour does play an important role in their reproductive behaviour, since in many species the production of scent, particularly amongst males, is related to sexual condition and the onset of maturity. It is likely that scent helps males and females of these predominantly solitary animals to locate each other during the breeding season.

The strong scent of mature males may function as an attractant to females and lessen their normally aggressive tendencies towards conspecifics. Mating in several species (including *Blarina*, *Cryptotis* and *Suncus*) has been observed to be preceded by much reciprocal sniffing. During courtship, the male repeatedly scent-marks the substratum and his pelage becomes increasingly impregnated with his own scent. This may prevent the female from attacking the male and induce receptivity. Shrews are induced ovulators, and the activity of the male is apparently necessary to make the female fully receptive. The length of time required for the sexes to consort before the female becomes fully receptive varies from less than an hour in *Sorex* species to one to two days in *Cryptotis*. Scent may play an important role in this behaviour.

It is not known if females in oestrus produce scent which indicates their condition, but this seems quite probable since males seem able to home in on sexually mature females. The absence of scent secretion from organs such as the flank glands in breeding females may encourage males to enter their home ranges or territories.

Conversely, the other main function of odour in shrews is to keep individuals apart, except during the breeding season when specialised scent glands come into operation. The generally strong smell exuded by shrews, together with scented faeces and urine, and the odours produced by an array of glands on the chin, neck, face and tail, serve to announce their presence. The scent can be very persistent: the pungent and distinctive odour of white-toothed shrews can be clearly detected several days after they have vacated a previously occupied area.

Odour may help individuals to orientate themselves within their home ranges, acting as a trail marker. It probably advertises residence in an area, and provides clear guidelines to other shrews about borderlines of home ranges and territories. As we have seen, studies of territoriality in shrews suggest that scented faeces act as a deterrent to invaders. Shrews sniff more but defecate and mark less when introduced to a cage occupied by a conspecific than to an unoccupied cage. Marking is frequently concentrated in the border areas between occupied areas. In these aggressive animals, such information probably assists mutual avoidance and reduces the number of direct confrontations.

SOUND

Shrews communicate more directly by means of calls and they produce a wide range of vocalisations serving a variety of purposes. Basically these calls fall into four categories according to the situation in which they are used: calls of alarm, defence and aggression; calls during courtship; vocalisations during interactions between a mother and her young; and sounds produced during exploration and foraging, and whose function is largely unknown. A summary of the vocal repertoire of shrews is given in Table 4.2.

Many of their vocalisations are high-pitched but still detectable to the

Table 4.2 Summary of the vocal repertoire of shrews

Description of sound	Context
Puts	Exploration/foraging; encounters with others
Clicks	Exploration/foraging; courtship
Twitters	Exploration/foraging; mother and young in the nest
Ultrasounds	Exploration/foraging
Short squeaks, chits	Alarm
Chirps, buzzes	Alarm; encounters at close quarters
Staccato shrieks	Aggressive encounters; warning
Rolling churls	Aggressive encounters at close quarters; warning; alarm
Chirps, twitters	Courtship
Twitters, clicks	Young in the nest
Squeaks, whistles	Young, hungry or cold, in the nest
Clicks, barks	Young outside nest eliciting attention

human ear, at least by young people whose hearing is good. As we age, our ears lose the capacity to detect sounds of the higher frequencies, and so the calls of shrews may be heard no more. A few shrew calls are in the ultrasonic range. Their hearing is acute and they respond most particularly to high-frequency noises: a loud low-frequency bang may elicit little reaction from a shrew, but a higher-frequency squeak has an immediate response. Captive shrews can be called out from their nests by soft, high-pitched squeaks or chirps.

In a study of captive *Suncus murinus*, *Blarina brevicauda* and *Cryptotis parva*, Gould (1969) could distinguish seven primary types of call in adult shrews, used in a variety of circumstances, and these were of low, intermediate or high intensity. Low-intensity sounds were produced in the range of 500–1,200Hz and within the bounds of human hearing. 'Puts' were low-frequency sounds produced by males and females and were used in encounters with other individuals when a series of these noises would be produced in quick succession. They were also employed when the shrews were exploring by themselves in a strange situation. They were very loud in *Suncus*, and detectable by other shrews up to about 60cm away. They were much softer in *Cryptotis* (a smaller species) and detectable only about 5cm away. Distinctive 'clicks' were emitted, again during solitary exploration of a strange place, and also in courtship behaviour between the sexes. A very common sound produced by many shrews, including those studies by Gould, is audible but soft, high-pitched 'twitters'. These appear to have no discernible function in direct communication but are produced almost continuously as individuals explore and forage, particularly in novel situations. They could serve as an advanced warning of approach to other shrews, but their precise function is unknown. Similar sounds are used between female *Sorex araneus* and their young in the nest. *S. minutus* is a much less vocal species and does not produce these audible twitters.

Sorex araneus and *Neomys fodiens* produce a succession of piercing, staccato squeaks or shrieks of warning when angry or alarmed, and these are the sounds most often detected by the human ear during spring and summer when shrews are actively searching for mates or meeting while crossing each others' territories. They can easily be heard along hedge-rows in quiet country lanes. At closer quarters, these shrieks may be replaced by lower-pitched rolling 'churls'. As two shrews meet, they freeze momentarily and then commence loud squeaking as they face each other. With their heads up, mouths open and snouts contracted, each one calls loudly to warn off the other. This vocal sparring may be sufficient to discourage further social intercourse and one of the shrews will then retreat.

S. minutus does not produce these loud shrieks and is much less vocal, merely emitting a single, short, audible 'chit' sound when threatened or alarmed. Upon meeting another conspecific, this sound is emitted and then the two shrews immediately move off in different directions. Single, sharp metallic squeaks are emitted by *Crocidura suaveolens* and *C. russula* under similar circumstances. *C. flavescens* produces loud, strident squeaks when confronting a stranger, plus a rattling 'chirr' when fighting. Soft 'chittering' sounds, which may function as contact calls, are produced when pairs are nesting together, and by the female to her young as she enters the nest and suckles them.

Medium-intensity calls of *Blarina*, *Suncus murinus* and *Cryptotis* according to Gould (1969), included an ultrasonic train of pulses of 50–107kHz which again were used during intensive investigation of objects but not in interactions with other shrews, and Gould speculated that these may function in echolocation (about which more will be said below). He also distinguished medium-intensity calls which were 'chirps' or 'twitters' produced by females when receptive to a courting male. Similar sounds are used by *Crocidura suaveolens*, *C. russula* and *Suncus etruscus*. The female of *Neomys fodiens* will also produce receptive 'chirps' when she meets a male, and while he follows her he emits a series of pure tone sounds during courtship. These sounds are presumed to signal the willingness of the female to mate and reduce her aggressive-ness towards the male.

High-intensity calls include 'chirps' and 'buzzes', in *Suncus murinus* and *Blarina* at least. 'Chirps' are produced when the shrews are suddenly frightened, and during encounters with other individuals at close quar-ters. They usually repel an approaching animal, and so may function to discourage the advance of conspecifics or predators. 'Buzzes' are produced in similar circumstances, again mainly in close encounters, and they may be emitted together with 'chirps'. The female is particularly vocal towards the male when she is unreceptive.

Infant shrews also produce a range of sounds, depending upon the circumstances. Warm, well-fed infants of *Crocidura flavescens* produce soft 'twitterings' and 'clicks', whereas hungry or cold infants utter high-pitched squeaks. When displaced from the nest, young *Blarina* emit 'clicks' which elicit a retrieval response from the mother. Infants of

1. *Common shrew (*Sorex araneus*) attacking an earthworm. (David Hosking)*

2. *Pygmy shrew (*Sorex minutus*). (David Hosking)*

3. European water shrew (Neomys fodiens) showing typical colouring including the white ear tuft and white fur behind the eye. (David Hosking)

4. European water shrew (Neomys fodiens) swimming underwater. Air trapped in the pelage creates a silvery appearance to the body when the shrew is submerged. (David Hosking)

5. *European water shrew (Neomys fodiens) grooming after swimming. This individual is the less common dark morph, lacking the white underparts typical of the species. (David Hosking)*

6. *Lesser white-toothed shrew* (Crocidura suaveolens). *Note the large ears and the long bristle-like hairs on the tail in this genus of shrews.* (David Hosking)

7. *Bi-coloured shrew* (Crocidura leucodon), *showing the large, naked ears typical of this genus of shrews.* (Frank Lane Picture Agency)

8. *Common shrew* (Sorex araneus). *Note the clear distinction between the dark fur of the back and the pale fur of the undersurface in this species. The flank gland is just visible as a small oval area amongst the pale fur between the fore and hind legs. (David Hosking)*

9. *Mature female common shrew* (Sorex araneus) *showing a mating mark on its head (see text). (David Hosking)*

10. *Close-up of the head of a mature female common shrew* (Sorex araneus) *to show the mating mark (see text). (David Hosking)*

11. *A nest of young common shrews (*Sorex araneus*) eight days after birth. (David Hosking)*

12. *A young common shrew (*Sorex araneus*) eight days after birth. It is naked and the eyes are not yet open. (David Hosking)*

*13. A young common shrew (*Sorex araneus*). (David Hosking)*

*14. American short-tailed shrew (*Blarina brevicauda*) eating a cricket. Note the very small ears and eyes, and the rather mole-like appearance of this semi-fossorial shrew. (Frank Lane Picture Agency)*

*15. European water shrew (*Neomys fodiens*) showing black and white pelage including white ear tuft. The stiff hairs on the edges of the toes, which assist propulsion when swimming, are just visible. (David Hosking)*

*16. A young common shrew (*Sorex araneus*).
(David Hosking)*

*17. Lesser white-toothed shrew (*Crocidura
suaveolens*) showing its teeth. (Frank Lane
Picture Agency)*

*18. Common shrew (*Sorex araneus*). (David Hosking)*

*19. Owls are major predators of shrews: here a barn owl has caught a common shrew.
(David Hosking)*

Suncus murinus emit 'whistles' under similar circumstances. The young of *Sorex araneus*, when a few days old, are capable of uttering loud 'barks' if hungry or displaced from the nest, and these also produce an immediate response from the mother. It seems that although distress calls of different species of young shrews sound different to the human ear, they will elicit the same response from adult female shrews. For example, the cries of abandoned infant *Crocidura suaveolens* produced immediate retrieval behaviour by a captive mother *Sorex araneus* who searched for them and carried them back to her own nest. Newborn *Sorex minutus* emit both contact whistles and distress cries during the first few days after birth. Later the infants, and the mother, utter a whispering sound when in the nest which may function as a recognition signal.

Those shrews which are capable of entering some degree of torpor, including *Crocidura russula* and *Suncus etruscus*, produce harsh shrieking calls if disturbed which, like the normal defence calls, have a tremolo structure. These calls seem to function as a means of defence to warn off intruders until the shrew has woken up sufficiently to be able to run away and hide.

The different sounds emitted by shrews are produced in three ways. Some, like 'puts' or 'hisses', are coupled with exhalation or inhalation through the nose. Others derive from clicking the tongue, and the remainder probably originate in the larynx, including the many 'chirps', 'twitters' and 'shrieks'.

Of particular interest are the high-pitched 'twitters' and other sounds produced by many shrew species as they explore novel situations. They tend to be emitted less frequently as they become more familiar with their surroundings. The components of some of these sounds are ultrasonic in frequency, and it has been suggested that echolocation may be employed to help shrews find their way around, and even in communication with their fellows. Since they have very poor eyesight, this could be a very valuable asset in various ways such as communication, detecting predators or food and exploring new areas. It may assist them to function effectively in darkness.

There is some evidence of echolocation in certain species which have been studied, namely the masked shrew (*Sorex cinereus*), the wandering shrew (*S. vagrans*), the American water shrew (*S. palustris*) and the American short-tailed shrew (*Blarina brevicauda*). These shrews have been found to emit high-frequency pulses from the mouth in the order of 30–60kHz of short duration (5–33ms or so) when exposed to strange surroundings. A similar system seems to occur in *Suncus murinus* and *Cryptotis parva*. Although echolocation is largely associated with bats, it is found in certain other groups of mammals such as whales. Various insectivores and tenrecs have been found to possess this faculty. Even blind humans have been reported to use it with surprising efficiency.

Among shrews, echolocation has been investigated in most detail in *Sorex vagrans*, the wandering shrew from North America. Buchler (1976) trained individuals of this species to locate a small platform about 11cm square situated some 7cm below an elevated disc on which they were placed, using only echolocation. They had to jump down on to this little platform and then run along a narrow ramp to a box where they would be rewarded with a morsel of food. Buchler took care to preclude all other cues such as olfactory and visual ones which might be used by the shrews in their task. He discovered that the shrews indeed uttered ultrasounds, and that these sounds were preferentially directed towards the little platform below them. When their ears were plugged with wax, their ability to locate the platform was significantly reduced, even though they increased their rate of ultrasound emission. Moreover, they were more hesitant about exploring when their ears were plugged. He found that these shrews would use echolocation when deprived of their kinesthetic memory and any other sensory information needed for orientation.

However, their echolocating abilities were very limited. Shrews were unable to detect a highly reflective, flat surface of 15cm by 15cm in area beyond a distance of 65cm, and thus were incapable of detecting the approach of a quiet predator. They may be able to identify the approach of a predator by listening to any ultrasounds produced by it, but when in familiar surroundings the shrews themselves produced ultrasounds quite infrequently. The smallest surface detectable at a distance of 20cm was an aluminium plate (highly sound reflective) of 3.5 by 3.5cm in area, so echolocation seems to be of very limited use in these shrews. However, it may provide them with useful information about their environment. They can, for instance, distinguish surfaces containing holes from solid surfaces so they may gain a so-called 'acoustic image' of their surroundings, including the whereabouts of gaps in the vegetation and burrow entrances which may save them the time and effort of tactile exploration.

The ultrasounds produced by S. vagrans are double pulses of 18–60kHz, together lasting about 8ms. Compared with the ultrasonic emissions by bats, those of shrews are much less intense. If shrews were to use high-intensity emissions in their cluttered habitats of plant stems, roots, logs and rocks they would be inundated with echoes from every conceivable surface. So high-frequency but low-intensity pulses are probably more useful. These low-intensity ultrasounds also attenuate rapidly and are easily scattered by small objects. This means that predators, even if they can detect ultrasounds, would have great difficulty locating the source of the sound. So echolocation in shrews may be a means of exploring the habitat without attracting the attention of predators.

Tomasi (1979) also suggested that echolocation in Blarina brevicauda was most useful in exploring the habitat, particularly the tunnel systems, searching for cover and avoiding obstacles when in a hurry. He found that these shrews produce ultrasonic clicks of 30–50kHz. Their echolocating abilities were rather better than those of Sorex vagrans, for they could distinguish between open and closed tubes, simulating burrow entrances, at a distance of up to 61cm. At 30.5cm they could locate openings as small as 0.53cm in diameter and corners up to 90° in angle.

More recently, evidence of echolocation has been found in the common shrew (Sorex araneus) by Forsman and Malmquist (1988). As with the experiments on S. vagrans, they found that S. araneus could distinguish between open and closed tubes. The shrews were kept within a central chamber which opened into six cylindrical tubes, each 200mm in length and approximately 16mm in diameter. The tubes were equidistant from each other, radiating out from the periphery of the central chamber.

The far ends of the tubes could be closed or opened at will by the experimenters who wished to find out if the shrews could distinguish one open-ended tube (from which they could then escape) from the five closed tubes by a process of echolocation. The experiments were conducted in darkness within a sound-proof box in order to eliminate orientation by sight or external sound.

The shrews soon adapted to the experimental conditions and rapidly effected their escape from the chamber via the open-ended tube, in most cases taking between 30 and 90 seconds to do so. They were able to discriminate between open and closed tubes at a distnce of 200mm, and to do so much more reliably than would be expected purely by chance. Ultrasonic vocalisations produced by the shrews were recorded during the experiment. These took the form of short-duration clicks with a broad sound range: some individual clicks ranged from around 25 to 95kHz. From this, Forsman and Malmquist concluded that shrews produced their own ultrasonic signals to use in echolocation. They also came to the conclusion that echolocation is used primarily during exploration of the environment.

While several of the vocalisations made by shrews during interactions with conspecifics have ultrasonic components, the extent to which they use ultrasounds in communication is not known. Rodents have been found to use ultrasounds in the context of the nest, as a means of communication between infants and mother, but there is no evidence of this in shrews.

VISION

The retina of shrews is well developed but because the eyeball is so small, sharp images are probably not formed and so vision is not an important means of communication, except perhaps at close quarters. Placing an object within centimetres of the face of a shrew commonly produces no reaction while a shadow falling over its body elicits a flight reaction. So sight may be useful primarily in detecting differences between light and shade. In the cluttered habitat of dense vegetation at ground level, vision has limited use anyway for an animal as small as a shrew, either in communication or in exploration. A captive, blind water shrew was reported to be able to find its way quite satisfactorily along familiar pathways.

Nevertheless, vision may play a significant role in communication during encounters between individuals at close quarters during terri-torial disputes and courtship. While no specific postures have been observed to accompany courtship behaviour, several different postures have been described when fights or scuffles break out between indivi-duals. These postures are described more fully below and include the stiff, elongated freeze posture which shrews usually adopt when they are aware of a conspecific's presence close by, the exposure of the neck and flank during some fights, and foot-stamping behaviour.

TOUCH

Since shrews have little contact with their fellows, except during the breeding season, tactile communication seems to be very limited indeed. It is probably of greatest use in the nest, between the mother and her infants. However, touch is an important method by which shrews can orientate and gain information about their surroundings during exploratory and foraging activities, using their long, sensitive and mobile snouts and their touch-sensitive vibrissae. There are several accounts of shrews bumping into a new obstacle placed on a familiar pathway, then quickly learning to avoid it by making a detour, and continuing to make the detour for a time after the obstacle has been removed. So not only do they orientate by tactile, auditory (echolocation) and olfactory cues, they also seem to employ both a spatial and kinesthetic memory of their home ranges and territories.

AGGRESSIVE POSTURES AND FIGHTING

Encounters between two strangers lead to a highly stereotyped repertoire of postures which are similar in all species of shrew, and exemplified by the common shrew (*Sorex araneus*). Encounters may be avoided by each animal keeping out of the other's way. If they meet unexpectedly, they usually utter a shriek of alarm and quickly move off in different directions. Sometimes, neither shrew will wish to retreat and as they meet they momentarily freeze, adopting a stiff posture with body outstretched and snout pointing forwards. After a moment of assessment, a shouting match ensues, with heads held high and loud, staccato shrieks being produced. If neither shrew retreats at this point, there will be a fight.

The dominant animal, usually the resident, may stamp its front feet on the ground and then attack the intruder. They rise on to their hind legs and lash out at each other with their front feet while still squeaking loudly. One may throw itself on its back, while squeaking and kicking, whereupon the other may run off. If neither retreats, they then aim bites at each other's heads and may become locked in combat, kicking and biting at each other while they roll on the ground, uttering low rolling churls as they do so. Eventually, one breaks free and rushes off into the undergrowth. It may be pursued by the opponent who bites at its rump and sides as it retreats. Shrews in agonistic postures are shown in Figure 4.1.

Encounters between *Crocidura flavescens* have two levels of aggression. These shrews appear to be territorial and resent the presence of strangers, although male and female home ranges may partially overlap. Low-intensity aggression often occurs between established pairs or individuals familiar with each other, and lasts only a few minutes. The two shrews face each other, squeak softly and occasionally lunge at each other, but never actually bite or indulge in a full-scale fight. High-intensity aggression usually occurs between two strangers. They also squeak and lunge at each other, but bites aimed at the muzzle of the opponent often make contact and may draw blood. One may then retreat,

Figure 4.1 Agonistic postures of shrews

pursued by the other who bites at its tail until it escapes into the undergrowth. The encounter is often followed by vigorous scent-marking, usually by the resident or dominant animal. If neither retreats, a full-scale fight ensues. The two rear up on their hind legs and box with their fore feet while aiming bites at each other. One may leap hind feet first towards the opponent in an attempt to knock it over. If this does not succeed, then a tumbling fight ensues. Such an encounter may last 40 minutes until one animal finally breaks away and retreats.

If two strangers are housed together, these encounters may occur intermittently for three or four days. But after that the two settle down and may even nest together.

Some of the postures typical of encounters between shrews may often be used in novel situations or when a strange object is discovered during normal exploratory activities. The elongated posture with head held outstretched and tail held out stiffly behind is used in response to the presence of a strange object as well as a conspecific. Tail-thrashing, which is exhibited by a receptive female to a courting male or by a subordinate male in an encounter with a dominant animal, is also a general sign of agitation. Shrews often thrash their tails vigorously from side to side when they are being handled following capture. Testing the air with raised snout and rearing up on the hind legs, often accompanied by soft twitters, also occurs in novel situations. In species such as *Blarina brevicauda* and *Suncus murinus*, marking behaviour, with the neck, flanks and belly being rubbed over the substratum, occurs following an encounter with another shrew, and also as an animal becomes accustomed to a new area.

5 Food and foraging behaviour

Since shrews devote most of their time to foraging, it is appropriate to consider this aspect of their natural history in some detail.

THE DIETS OF SHREWS

Shrews are essentially opportunistic predators which feed on a wide variety of common invertebrates, particularly beetles, bugs, earthworms, woodlice, spiders, slugs, snails and insect larvae. The scale of diversity of their diets is indicated by the many detailed studies of their feeding habits. For example, stomach analyses revealed some 25 different prey taxa in the diet of the pygmy shrew (*Sorex minutus*) in Ireland studied by Grainger and Fairley (1978). I found a similar number of prey types was found in the diet of the common shrew (*Sorex araneus*) in southern England by analysis of faecal pellets of live-trapped animals (Churchfield, 1984b). Forty-two prey taxa were identified as food of *Sorex cinereus*, the masked shrew, in Indiana, USA by French (1984). By exploiting both terrestrial and freshwater prey, the semi-aquatic water shrew (*Neomys fodiens*) was found to feed on some 36 different prey taxa (Churchfield, 1984b). Even littoral prey are taken: the Scilly shrew (*Crocidura suaveolens*) commonly occurs in coastal habitats including the boulder zone on the seashore where it forages for amphipod crustaceans as well as other small invertebrates associated with the strand-line. Most invertebrates, therefore, fall prey to shrews. Examples of the range of prey items regularly taken by three species of shrew in England are shown in Table 5.1.

While most shrew species take a wide range of invertebrate prey, this does not necessarily imply a complete lack of food choice amongst them. They do discriminate between prey and show preference for some and distaste for others. The common shrew (*Sorex araneus*), for example, exhibits choice over the species of terrestrial woodlice eaten: *Philoscia* is eaten readily while *Armadillidium*, with its very thick exoskeleton and its ability to roll into a ball, is mostly ignored (Crowcroft, 1957). Millipedes are rarely eaten, despite being common litter invertebrates, and yet centipedes are often taken. Rudge (1968) suggested that the acrid

Table 5.1 The percentage frequency of occurrence of different prey items found in the diets of shrews in southern England (after Churchfield, 1984b)

		Sorex minutus (pygmy shrew)	*Sorex araneus* (common shrew)	*Neomys fodiens* (water shrew)
		---	---	---
	No. samples	35	259	161
	Terrestrial Prey			
BEETLES	Carabidae	5.7	5.0	6.8
	Staphylinidae	8.6	7.7	9.9
	Chrysomelidae	8.6	9.3	1.9
	Other Coleoptera adults	48.6	18.1	18.0
	Coleoptera larvae	0.0	6.6	3.1
BUGS	Hemiptera	25.7	12.4	8.7
FLIES	Tipulidae larvae	0.0	7.3	2.5
	Other Diptera larvae	0.0	10.8	9.3
BUTTERFLIES	Lepidoptera larvae	11.4	6.2	2.5
ANTS	Formicidae	8.6	10.4	7.5
	Other Hymenoptera	0.0	0.8	0.6
EARWIGS	Dermaptera	0.0	0.4	0.6
SPRINGTAILS	Collembola	5.7	1.2	0.6
MITES	Acari	25.7	25.1	11.8
SPIDERS	Araneae	60.0	20.5	12.4
HARVESTMEN	Opiliones	17.1	10.4	5.0
WOODLICE	Isopoda	51.4	30.5	9.3
CENTIPEDES	Geophilomorpha	5.7	13.9	6.8
	Lithobiomorpha	11.4	2.3	1.9
MILLIPEDES	Diplopoda	0.0	0.0	6.8
SNAILS/SLUGS	Gastropoda	14.3	53.7	21.1
EARTHWORMS	Lumbricidae	0.0	63.7	21.1
	Aquatic prey			
BEETLES	Coleoptera adults	0.0	0.0	3.7
	Coleoptera larvae	0.0	0.0	0.6
BUGS	Hemiptera	0.0	0.0	0.6
CADDIS FLIES	Trichoptera adults	0.0	0.0	1.2
	Trichoptera larvae	0.0	1.2	41.0
STONEFLIES	Plecoptera larvae	0.0	1.2	16.8
MAYFLIES	Ephemeroptera nymphs	0.0	0.0	1.9
FLIES	*Simulium* larvae	0.0	0.0	8.7
	Other Diptera larvae	0.0	3.1	24.2
WATER SLATERS	*Asellus*	0.0	0.4	70.2
SHRIMPS	*Gammarus*	0.0	0.0	26.7
SNAILS	Gastropoda	0.0	0.0	1.2
OSTRACODS	Ostracoda	0.0	0.0	7.5
FISH	Osteichthyes	0.0	0.0	1.2

secretions produced by some prey, such as the millipede *Ophyiulus pilosus* and the snail *Oxychilus alliarius* made them unpalatable to shrews so that although they were occasionally eaten in laboratory trials they were rarely taken when other prey were available. Certain species of shrew have notable omissions from their diets. Earthworms are a major prey of many shrews but they rarely, if ever, feature in the diet of pygmy shrews (*Sorex minutus*); there is only one report of them doing so (Churchfield and Brown, 1987). When provided with even quite small earthworms, captive pygmy shrews ignore them. The reason for this may be twofold: they may find that earthworms are generally too large and difficult to handle or, being typically foragers on the ground surface, they may not encounter these prey as often as the more subterranean common shrew (*S. araneus*).

The size of invertebrate prey taken by shrews ranges from tiny collembolans (springtails), aphids and mites of 3mm or less in body length to large earthworms of 60mm or more. However, the extremely small size of some prey such as mites cannot preclude the possibility of accidential ingestion, although Shillito (1960) found that common shrews consistently ate springtails found amongst leaf litter. It seems that all shrew species exploit the full range of prey sizes available. For example, Figure 5.11 shows the dietary composition of *Neomys fodiens*. *Sorex araneus* and *S. minutus* with respect to the prey sizes eaten. Even the large *N. fodiens* takes some very small prey, and the diminutive *S. minutus* takes some large prey. All three species took most of their prey in the range of 6–10mm in body length.

While the main bulk of shrews' diets comprise small invertebrates, there are many examples of vertebrate prey being caught and eaten. While several terrestrial shrews such as *Crocidura russula* and *Blarina brevicauda* are known to catch lizards and small mammals, it is the semi-aquatic *Neomys fodiens* which makes the most use of vertebrate prey by catching frogs, newts and small fish (Wołk, 1976). The occurrence of these large prey in the diets of shrews is often, but not exclusively, associated with the possession of venomous saliva which helps to immobilise the prey, and this will be returned to later.

Some shrews have been reported to scavenge occasionally, and vertebrate or other carrion may feature in the diet. Schlueter (1980) suggested that carrion eating by *Sorex araneus* and *Neomys fodiens* may be most important in winter. Shrews feed primarily on animal food but there are many reports of plant material being eaten, usually only in small amounts. The guts of shrews have often been found to contain small fragments of grasses and herbs. (e.g. Bever, 1983), even considerable quantities of seeds and, again, this may have a seasonal bias with larger amounts being taken in winter, possibly to compensate for poor availability of invertebrate prey (Lavrov, 1943; Kangur, 1954). The impact of their seed-eating habits on the germination success of trees has even been speculated upon (Moore, 1942).

It is clear, then, that shrews are generalists in terms of the spectrum of food types taken, and yet some degree of specialisation does occur with respect to the proportions of the different food items featuring in their diets. There is a quantitative rather than a qualitative specialisation which is apparent in those species whose diets have been studied in most

*Figure 5.1 Dietary composition of common, pygmy and water shrews (*Sorex araneus*), S. minutus and Neomys fodiens) showing the major prey types eaten. Each block represents a maximum of 30 per cent by composition (after Churchfield, 1984b)*

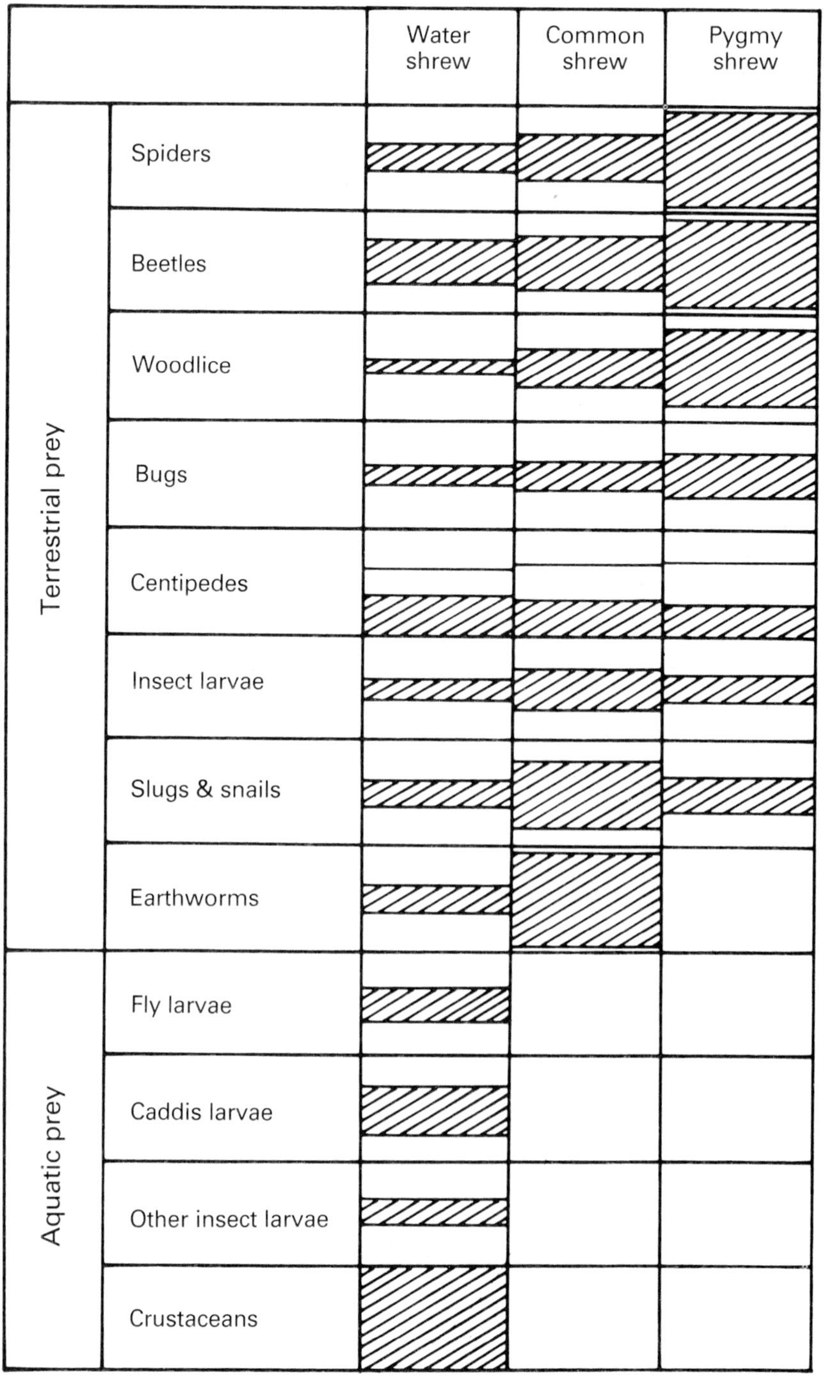

detail. Figure 5.1 shows the diets of three common European shrew species occurring in the same habitat in southern England. While a variety of prey was eaten, the bulk of the diet of all three species comprised only three prey types. In the case of *Sorex araneus*, these were earthworms, gastropods and coleopterans, which together constituted over 50 per cent of the diet. In *S. minutus* they were araneids, coleopterans and isopods, and in the semi-aquatic *Neomys fodiens*, they were aquatic crustaceans, trichopteran larvae and coleopterans. Similarly, only five major prey types featured in the diets of the white-toothed shrew, *Crocidura russula* (Bever, 1983), and two North American shrews, *Sorex longirostris* and *S. cinereus* (French, 1984).

DIETS AND PREY AVAILABILITY

The diets of shrews appear to be a reflection of what is available as prey, although the extent to which this holds true has not been fully established. The very fact that shrews exhibit high dietary diversity coupled with generalist and opportunist habits is evidence of an adaptation to overcome spatial or temporal shortages of particular prey types. The relationship between diet and food availability in shrews has been the subject of several studies. Rudge (1965) found a close correlation between the abundance of certain potential prey items in the leaf litter and their occurrence in the diet of *Sorex araneus*. Major groups, such as lumbricids, isopods and gastropods, which had a high frequency of occurrence in shrew guts and a high individual biomass, tended to be prominent in the diet of *S. araneus* in all months, despite changes in their availability. But the occurrence of secondary groups, such as centipedes, spiders and insect larvae, was more closely dependent upon their current status in the litter.

I found little correlation between the diet of *S. araneus* and seasonal numbers of most prey types (Churchfield, 1982a). Some prey, such as earthworms, were occasionally present in considerable numbers in the habitat and yet failed to appear in the diet. Conversely, some (including insect larvae) occurred more frequently in the diet than I would have expected from their low numerical availability. Only with coleopterans was there a clear relationship between their occurrence in the diet and their abundance in the habitat (see Figure 5.2). An increase in the numbers of beetles available was accompanied by an increase in their predation by shrews.

Pernetta (1976) was unable to find good positive correlations between the diet of *S. araneus* and the numbers of different prey types collected in pitfall traps, but he did get close correlations between the diet of *S. minutus* and certain prey, namely spiders, beetles and insect larvae found active on the ground surface.

So while there is some relationship between diet and food availability, shrews are not wholly opportunist predators for they exhibit selection for certain prey regardless of their abundance.

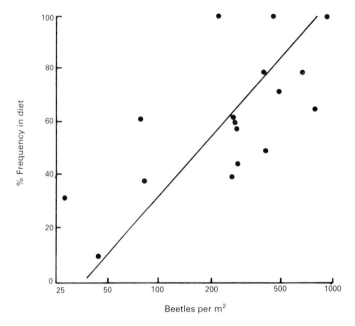

*Figure 5.2 The relationship between the abundance of beetles and their occurrence in the diet of the common shrew (*Sorex araneus*) (after Churchfield 1982a)*

FOOD CONSUMPTION

Shrews are renowned for their voracious appetites. For example, *Sorex araneus* requires 80–90 per cent of its body weight in food daily or 6.7–9.7KJ/g/day and the smaller *S. minutus* requires 125 per cent of its body weight daily or 9.7–13.0KJ/g/day (Hawkins and Jewell, 1962; Churchfield, 1979). What this means in terms of the number of prey which must be captured in order to survive from day to day is difficult to estimate because of the great range in size and food value of the prey taken. However, a shrew of average size, such as *Sorex araneus* weighing 8g, requires about 100 prey of approximately 10mm in body length per day.

For many species of shrew, food consumption is related to body weight. Larger shrews require more food in terms of absolute amounts but less food as a percentage of their body mass than small shrews, as Figures 5.3 and 5.4 indicate. However, food consumption also depends upon the type of shrew, particularly its phylogeny and metabolic rate. Shrews fall into two quite distinct subfamilies reflecting different phylogenetic or evolutionary lines: the Soricinae and the Crocidurinae. Most northern, temperate species belong to the former, including the familiar red-toothed shrews of the genus *Sorex*. These shrews have characteristically high metabolic rates and food consumptions. The Corcidurinae, on the other hand, contains a multitude of species including the many white-toothed shrews (genus *Crocidura*) which orginated from more tropical climes and have much lower metabolic rates (Vogel, 1976). *C. russula*, for instance, requires only about 48 per cent of its body weight daily or

94

3.4KJ/g/day (Churchfield, 1979). The relatively high food intake of shrews can also be attributed to the high water content of prey and the quantity of indigestible chitinous exoskeleton ingested.

The energetics of shrews are discussed more fully in Chapter 6. High energy requirements have an important bearing on the dietary composition and the foraging strategies adopted by these highly active mammals, particularly their need for frequent foraging bouts and the decisions made concerning prey choice and food availability.

Figure 5.3 The relationship between absolute food consumption and body mass of shrews

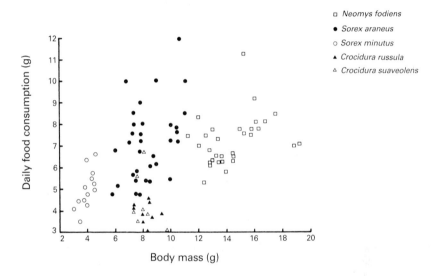

Figure 5.4 The relationship between food intake as a percentage of body mass and the size of shrews

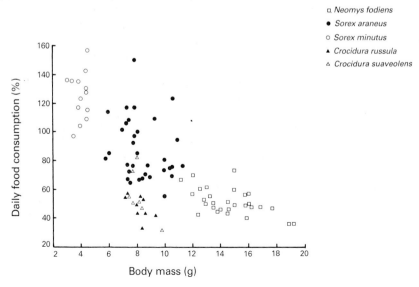

PREY DETECTION AND HANDLING

The Role of the Senses in Prey Detection

The olfactory, auditory and tactile senses of shrews are generally well developed but sight is poor. The eyes of shrews are so small that their use would appear to be very limited, as we have seen in Chapter 1. The prime function of vision in most shrews seems to be simply to record light intensity, for Branis (1981) could not observe any discernible reactions of shrews to light stimuli, or to moving stimuli (including prey) unless another sense such as touch, hearing or olfaction had been stimulated. Rood (1958), Vlasak (1970) and Branis (1981) all concluded that sight is of little use in prey detection.

The exact role of scent in prey detection by shrews is difficult to assess. They certainly have a well-developed sense of smell which is evident not only from the large olfactory lobes of the brain and the complex arrangement of turbinals in the snout, but also by the common use of scent in their social life. Holling (1958), Pernetta (1977b), Schmidt (1979) and Churchfield (1980b) all concluded that olfaction does have an important function in prey detection. I demonstrated the prodigious abilities of captive common shrews (*Sorex araneus*) to locate and dig up insect pupae buried at different depths in the soil in outdoor enclosures. Although the success rate of locating prey declined as depth at which they were buried increased, shrews exhibited a 50 per cent recovery rate of groups of five pupae at 120mm depth and an 8 per cent recovery rate at 160mm depth. Pupae buried singly were not recovered with such ease, yet even then there was a 30 per cent recovery rate at 120mm, although none were found at 160mm. However, it seems that olfaction may be a useful but not essential cue to prey detection, since pupae whose scent was masked by a coating of varnish unfamiliar to shrews were recovered in equal proportions to unvarnished pupae at 50mm depth. Olfaction may be more important in determining the quality of food once it is located for Holling (1955) found that shrews were able to distinguish between healthy pine sawfly cocoons and those parasitised by fungus, and reject the latter before opening them.

Shrews may differ in their olfactory abilities according to their mode of foraging. For example, Pernetta (1977b) found that pygmy shrews (*Sorex minutus*) were less efficient at detecting buried prey items by smell than either common shrews (*S. araneus*) or white-toothed shrews (e.g. *Crocidura suaveolens*). This he correlated with the tendency for the pygmy shrew to forage on the ground surface while the other species are more subterranean in habits.

Auditory cues also seem to assist shrews in prey detection (Pernetta, 1977b). For instance, rustling noises will elicit an immediate reaction from captive shrews who will investigate the source of the sound. A captive pygmy shrew kept by the author showed great excitement when a large blowfly was confined in its cage. The shrew chased the prey as it flew around the cage, jumping at it until the fly was knocked to the floor and captured. It has been speculated that auditory perception of prey may even extend to the use of echolocation in some shrews. For example, the echolocating abilities of the wandering shrew, *Sorex vagrans*, (Buchler, 1976) and the short-tailed shrew, *Blarina brevicauda*, (Tomasi, 1979) have been demonstrated but they operate only crudely and over short ranges and so are unlikely to be of much use in the detection of prey, especially with all the background clutter of high-frequency noises at ground level in their natural habitats.

One of the most characteristic features of shrews is the long, sensitive snout which is well furnished with touch-sensitive whiskers or vibrissae, and Eimer's organs (Pernetta, 1977b). The vibrissae, whose follicles are well supplied with nerve endings and nerve fibres, may well relay tactile information concerning small movements of prey and yet it seems that they are by no means essential to the process of prey detection. For example, clipping the vibrissae did not apparently impair underwater prey detection by *Sorex palustris* (Sorenson, 1962) or *Neomys fodiens* (Churchfield, 1985) in any way.

Prey detection, then, seems to be aided by a combination of these senses, particularly smell, sound and touch, none of which are very effective when used alone, coupled with random searching for prey. The long, mobile, cartilaginous snout is used to probe amongst vegetation, leaf litter and soil for food. Shrews do not apparently search systematically for prey, they simply move rapidly through the undergrowth or through a subterranean burrow, probing with the snout and sniffing until they are aware of the presence of prey. They then home in on it by moving rapidly to and fro, any movement of the prey stimulating attack. When a clump of prey, such as insect larvae or pupae, is located they will return to the same spot and search until the supply is exhausted (Churchfield, 1980b).

Prey Capture and Handling

Prey capture by shrews simply involves co-ordinated use of the mouth and fore feet. Prey are quickly pounced on when encountered. Larger prey are pinned to the ground with the fore feet and the teeth are used to disable them before they are rapidly eaten, being passed laterally between the two rows of teeth. Highly chitinised parts such as legs and wing cases are often rejected. Slugs may be eaten only once the slime has been rubbed off by the fore feet or against the substratum. Slugs also require a great deal of mastication, and the time taken to eat even a small slug is longer than that for an arthropod of similar size. Small prey are chewed up whole. However, some shrews possess poison glands producing narcotising saliva as an adjunct to prey handling, a rare phenomenon amongst mammals.

Two shrews are particularly well known for their venomous saliva: the American short-tailed shrew, *Blarina brevicauda* (Figure 5.5), and the European water shrew, *Neomys fodiens*. The first lower incisors of both these shrews have concave medial surfaces forming a crude channel. At the base of these teeth are the openings to the ducts from the toxin-producing submaxillary glands. As the shrew bites its prey, the toxin, mixed with saliva, is transported from the glands into the prey via the incisor teeth: *N. fodiens*, at least, is reported to salivate copiously as it attacks its prey.

Extracts of the toxin produced by these shrews affect the nervous,

*Figure 5.5 A short-tailed shrew (*Blarina brevicauda*) killing a frog*

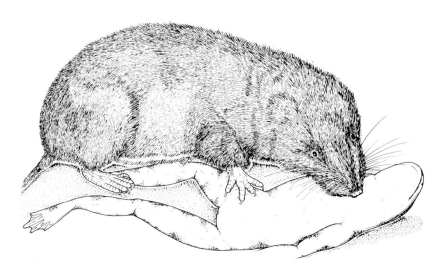

respiratory and vascular systems causing convulsions, unco-ordinated movement and eventually paralysis and death in small vertebrates such as frogs and mice (Pucek, 1959, 1967). A bite from *N. fodiens* will cause reddening of the skin and some discomfort at the site of the bite, which may persist for several days, in human subjects. The strongest toxic action has been demonstrated in *B. brevicauda*. *Neomys anomalus* has also been shown to produce toxic saliva, but its effects are much milder (Pucek, 1969).

N. fodiens feeds on fish and frogs, and *B. brevicauda* can kill small mammals considerably larger than itself. Both these shrews tend to attack large prey from behind, directing bites at the neck where the neurotoxin might most readily be injected into the central nervous system. Eadie (1952) reported that field voles (*Microtus pennsylvanicus*) are an important autumn and winter food for *B. brevicauda*. The ability to employ a narcotising saliva may, therefore, be of considerable adaptive importance to these shrews when wanting to tackle large prey.

Food-Hoarding

Many shrews, when maintained in capitivity, will readily hoard prey when presented with a superabundance of food, although the extent to which this occurs in the wild is not known. *Blarina brevicauda*, *Sorex araneus* and *Neomys fodiens* are particularly avid food-hoarders. *B. brevicauda*, for example, will cache a variety of food items including earthworms (Tomasi, 1978), snails (Ingram, 1942), insects (Martin, 1981) and even small mammals (Tomasi, 1978), sunflower seeds (Martin, 1984) and other plant material (Hamilton, 1930).

A study of caching behaviour in *Blarina brevicauda* by Robinson and Brodie (1982) revealed that shrews offered different prey species would systematically cache the majority of these items. Moreover, cached prey were marked with urine and faeces, and shrews would later return and consume them. A surplus of prey seems to stimulate caching behaviour, as in moles, for Buckner (1964) found that hoarding of larch sawfly cocoons increased as prey density increased. Caching also seems to be stimulated by changing environmental conditions. Martin (1984) showed that captive *B. brevicauda* under simulated natural conditions would commence food-hoarding in or near the nest in autumn, continue through winter and cease hoarding in spring, by which time all stores had been used up. Shrews confined in uniform, climate-controlled conditions did not make proper caches: although hoarding under summer conditions could be induced by providing shrews with a sudden abundance of prey, these stores lasted only a few days compared with the extended periods of the winter stores. Some prey were hoarded alive and yet immobilised, much as in the manner of moles. Martin concluded that season, prey abundance and possibly the proximity of conspecifics were the primary factors influencing food-hoarding by *B. brevicauda*.

The food caches of shrews are usually associated with their own nests but Platt (1976) found that the abandoned nests of field voles (*Microtus*) were adopted as major sites for caches by *B. brevicauda*. I noted that besides storing food in the nest, *S. araneus* kept in outdoor enclosures made additional caches some way from the nest by excavating shallow depressions in the soil in which food was deposited and covered over firmly with leaf litter (Churchfield, 1980b). *Neomys fodiens* is also reported to store food, but in secluded places under stones. Given the opportunity it seems that shrews, both in capitivity and in the wild, will make multiple caches. However, not all individuals of these different species show the hoarding habit. Those who maintain captive shrews well know that certain individuals will systematically cache the whole contents of their food pots whereas others show no inclination to do so.

The ability to cache unused prey ensures that a predictable energy source will be accessible as insurance against times of food scarcity or adverse conditions which may reduce foraging success. This could be of particular importance to the smaller temperate shrews which remain active throughout the winter. However, caching food also incurs a cost, for it has to be guarded from competitors who may steal it. Territorial behaviour may provide sufficient deterrence to would-be thieves, and make food-hoarding worthwhile. The fact that wild shrews are unable to store body fat in sufficient quantities to help them overwinter (Churchfield, 1981), may be partly compensated for by hoarding food.

FORAGING MODES AND STRATEGIES

Foraging Modes Adopted by Shrews

All shrews obtain much of their food by foraging on the ground surface where a process of random but thorough searching is used to locate food, largely with the aid of olfactory and tactile cues. However, they are not restricted to this mode of foraging for they have adopted other means of obtaining food, for which they are adapted to a greater or lesser extent. Three major foraging modes may be distinguished with respect to diet, habitat use and the degree of anatomical or physiological adaptation. These are epigeal, hypogeal and aquatic. In addition, Hutterer (1985) recognises several scansorial shrews (adapted for climbing) on the basis of anatomical features, but little is known of their habits. The foraging modes employed by different shrews are summarised in Table 5.2. Nevertheless, because all shrews remain essentially generalists in design, lacking the extremes of adaptation which would otherwise restrict their way of life, these categories are not mutually exclusive and species may indulge in more than one foraging mode. This is indicated in Figure 5.6 where the dietary composition of three shrew species coexisting in the same habitat is shown. While the common shrew (*Sorex araneus*) took mostly subterranean prey, it also foraged on the ground surface, and the water shrew (*Neomys fodiens*), although primarily aquatic in foraging mode also took prey from terrestrial locations.

Epigeal foragers are found primarily or exclusively on the ground surface where they forage for surface-dwelling invertebrates. Although

Table 5.2 Classification of shrews into different foraging modes based upon available information about anatomical adaptations and feeding habits. (?) Speculative

		Foraging mode			
Genus	Common name	Epigeal	Hypogeal	Scan-sorial	Aquatic
Blarina	Short-tailed shrews		4		
Blarinella	Asiatic short-tailed shrew		1		
Feroculus	Kelaart's long-clawed shrew		1		
Solisorex	Pearson's long-clawed shrew		1		
Anourosorex	Mole shrew		1		
Cryptotis	Small-eared shrew	4	8		
Myosorex	Mouse shrews; forest shrews	4	8		
Sylvisorex	Forest musk shrews	4	2	1	
Soriculus	Asiatic mountain shrews	6	1	3	
Sorex	Red-toothed or long-tailed shrews	55 (?)	3	3	3
Notiosorex	Desert shrews	1			
Suncus	Musk shrews	13		1	
Crocidura	White-toothed shrews	125 (?)			
Paracrocidura	Cameroon shrew	2			
Scutisorex	Armoured shrews	1			
Megasorex	Mexican shrew	1			
Diplomesodon	Turkestan desert shrew	1			
Neomys	Eurasian water shrews				3
Chimarrogale	Oriental water shrews				4
Nectogale	Tibetan water shrew				1
Total species		217	30	8	11

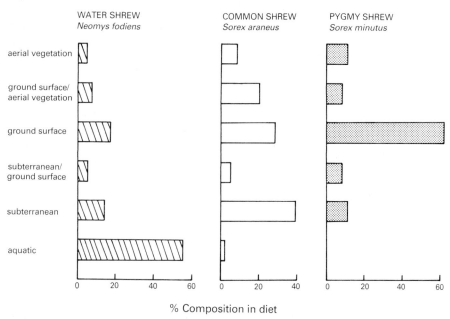

*Figure 5.6 Dietary composition of common, pygmy and water shrews (*Sorex araneus,
S. minutus *and* Neomys fodiens*) with respect to the location of prey (after Churchfield,
in press)*

they may probe amongst vegetation, leaf litter and surface soil for food,
they do not actively dig for prey. They include the smallest species such
as *Sorex minutus, S. minutissimus, S. hoyi* and *Suncus etruscus* which
are particularly light in build and lack the ability to burrow effectively,
although they are small enough to make use of existing cracks and
crevices.

There are no true shrews which are highly adapted for burrowing and a
subterranean existence to the extent of their relatives the moles. But
there are species which make frequent use of underground tunnels,
usually made by other mammals, or forage amongst soil and leaf litter in
search of soil invertebrates which comprise the major portion of their
diet. Hence these shrews are not really fossorial but are referred to as
being hypogeal (active below the ground surface). They are typically
thick-bodied and robust, and of medium or large size, which enables them
to push their way through soil and leaf litter, using the snout, teeth
and fore feet to excavate a passage where necessary. The skull may be
shorter and more heavily constructed and the digits of the fore feet rather
longer in those best at digging, such as *Anourosorex* (Figure 5.7) and
Diplomesodon. A wide variety of shrews habitually use this mode of
foraging, including *Sorex araneus, S. unguiculatus,* and *S. trowbridgii,
Blarina brevicauda* and species of *Cryptotis* and *Myosorex*. Typically,
earthworms and soil-dwelling insect larvae form a major food resource
for hypogeal shrews, which distinguishes them from the more epigeal
species, and there is some evidence to suggest that the former have a
greater olfactory ability to assist with the location of these hidden prey.
Rather little is known about the foraging strategies of these shrews.

*Figure 5.7 The mole shrew (*Anourosorex squamipes*) of eastern Asia which is adapted for a subterranean mode of life*

All shrews are accomplished swimmers and several species have adopted an aquatic mode of foraging whereby they hunt for prey underwater. But, again, most of these species have only rather rudimentary adaptations for this lifestyle. The aquatic habit occurs very diversely amongst shrew genera, providing a good example of convergent evolution. There are three species of Old World water shrews of the genus *Neomys* found in Europe and northern Asia, of which the most familiar is *N. fodiens*. Three species of the genus *Sorex* in North America are aquatic, *S. palustris* being the best known. There are four Asiatic water shrews (genus *Chimarrogale*) which are found in eastern and southeastern Asia, including the Himalayas, parts of China, Sumatra and Borneo. All these species possess little in the way of adaptations for swimming and diving beyond a better-insulated pelage than their terrestrial counterparts and fringes of stiff hairs on the feet to increase the surface area and assist swimming (see Figure 1.8). Only *Nectogale elegans*, the so-called elegant or Tibetan water shrew, shows a higher degree of specialisation for this mode of life: it is the only shrew to have webbed feet (Figure 5.8). Not only are the edges of the feet furnished with stiff hairs, as in the other species mentioned, but both the fore and hind feet have toes which are fully webbed. It is also reported to have teeth which are specialised for eating fish.

Most water shrews are neither anatomically or physiologically well adapted for diving. Of those species which have been studied, the dives are of relatively short duration, ranging from a mere four seconds in *Neomys fodiens* (Churchfield, 1985) to 48 seconds in *Sorex palustris* (Calder, 1969), and buoyancy is a major problem since the body is not very streamlined and the thick pelage traps air which may be a useful attribute to reduce wetting but which makes the animal even more buoyant. The air trapped between the hairs of the fur creates the appearance of a silvery film around the body when it is submerged, and this is often commented on by observers. The shrews frequently return to land and any prey caught is usually carried to land for consumption.

*Figure 5.8 The Tibetan water shrew (*Nectogale elegans*) which is the best adapted of shrews for swimming*

Water shrews feed on a variety of freshwater invertebrates and even frogs, newts and small fish. I found that the bulk of the diet of the European water shrew, *Neomys fodiens* comprised freshwater shrimps and other crustaceans such as *Gammarus* and *Asellus*, and caddis fly larvae, as portrayed in Figure 5.1 (Churchfield, 1984b). Underwater foraging occurred both in summer and winter: between 33 and 67 per cent of the diet was of aquatic origin, with a mean of 50 per cent over the whole year, and no significant seasonal trends were apparent. The remainder of the diet comprised land invertebrates such as are eaten by the more terrestrial shrews.

Aquatic foraging must be an energetically very costly strategy, particularly for animals in temperate climates, and it is interesting to note that water shrews are most commonly found associated with cold, swift-flowing mountain streams in which they forage all year round. *Sorex alaskanus* inhabits the coastal region of south-eastern Alaska, while *S. palustris* extends even further north in Alaska and Canada. *Chimarrogale* and *Nectogale* are found in the Himalayan region. Although the food supply may be more predictable than in terrestrial habitats which are subject to greater seasonal changes, nevertheless the success rate of dives is not guaranteed. Wołk (1976) found remains of small twigs, stones and other inanimate objects at feeding stations which *Neomys fodiens* had retrieved from the water, apparently mistaking them for prey. This was confirmed when I found that captive water shrews could be tricked into retrieving inanimate cylindrical objects from the water (Churchfield, 1985). It seems that the shape of the object rather than movement may be a more important rule-of-thumb guide to prey detection in aquatic shrews. The trade-off for aquatic foraging may be access to resources which provide a high calorific return and/or a means of avoiding interspecific competition with more terrestrial shrews.

However, none of these shrews appear to have vacated the terrestrial foraging niche altogether.

Foraging Strategies of Shrews

There has been increasing interest in the foraging strategies of shrews as a result of the developments in optimal foraging theory in recent years. Optimal foraging theory operates on the assumption that predators are subjected to choices in their daily routines, such as what kind of prey to take, where to search for food and when to move to a new feeding patch. Those choosing the most efficient or optimal solutions will be the most successful because daily decisions made about foraging strategies will contribute to an individual's chance of survival and reproduction. An efficient forager which is aware of the costs and benefits of different strategies and can act upon them will also have more time available for other pursuits such as finding a mate and rearing young. A number of predictions have been made in the context of optimal foraging theory, such as: predators should select more profitable prey, with respect to size, handling times and food value; they should feed more selectively when profitable prey are abundant; and they should ignore unprofitable prey, regardless of how numerous they are, when profitable prey are abundant.

While most studies of optimal foraging have been carried out with birds (e.g. Smith and Sweatman, 1974; Cowie, 1977; Krebs *et al.*, 1977, 1978) there is increasing interest in other vertebrates. Shrews make particularly good subjects for study because of their high energy requirements and the consequent need for frequent bouts of foraging. Their general lack of shyness in laboratory situations also means that they are relatively easy animals to observe. Barnard and Brown (1981) examined the theory that predators may sometimes use rules-of-thumb such as prey size to estimate relative prey profitability for optimising prey selection. They provided *Sorex araneus* with a choice between large and small pieces of mealworm distributed randomly in small wells on a grid, where large pieces were least profitable in terms of calorific value per unit handling time. They predicted that if shrews selected prey on the basis of profitability they should select small prey when abundant and ignore large prey. If they selected according to size, then they should prefer large prey. They found that shrews tended to prefer large prey (despite their lower relative profitability) but this depended upon their encounter rate with these prey. When encounter rates with large prey were low, shrews were equally likely to take large or small prey. When encounter rates with large prey were high, shrews were more likely to take large prey than small prey. Changes in encounter rates with small prey, once encounters with large prey exceeded a certain level, did not affect the shrews' preference for large prey. Hence, size was used as a guideline for prey choice, with shrews showing clear selection for large prey.

Barnard and Brown (1985a) also investigated the effects of a competitor on the selection of prey by test animals by introducing a second shrew to the experimental area. This shrew had not been trained to forage on the grid and so did not deplete the food supply. Prey were presented such that the encounter rate with large prey was high enough for them to selectively take large prey. The prediction that predators should become less selective in the presence of a competitor was born out as the degree of selection for large prey was significantly lower than that expected from the encounter rates with large prey.

Risk-sensitive foraging theory predicts that predators with frequent and high energy demands, which may starve if there is a temporary shortage of food, should choose their feeding areas on the basis of variation as well as mean expected reward rate. Thus, for a given mean reward rate a predator should be 'risk-prone' by selecting high-variance feeding areas if it is running below its energy requirements because it has a chance of finding an abundance of food and replenishing its losses quickly, but it should be 'risk-averse' by selecting low-variance feeding areas if it is keeping above its energy requirements.

Barnard and Brown (1985b) investigated the risk-sensitive foraging behaviour of common shrews (*Sorex araneus*) by presenting captive individuals with a choice between constant and variable-supply feeding stations. They found that shrews were indeed more likely to visit the variable supply station when running below energy requirements and the constant-supply station when running above. But the tendency towards risk-aversion above energy requirements was greater than that towards risk-proneness below requirements.

Whether insectivores such as shrews exhibit prey selection according to these theories of optimal foraging in the wild where they encounter prey of many types and sizes and live in the presence of other potential competitors of the same and different species seems unlikely. Field-based studies of the feeding habits of shrews suggest that they eat whatever they encounter and reject little, but recent research on niche overlap and possible modes of ecological separation may shed more light on this question.

REFECTION

Shrews sometimes indulge in a habit known as refection when they will curl up on their sides or backs, usually when in the nest, and spend a few moments licking up a milky, whitish fluid from the everted rectum (see Figure 5.9). This activity has been likened to the coprophagous feeding habits of rabbits and, although shrews do not appear actually to eat their faeces, refection may have a similar function in helping to extract further nutrients from the food. Although it has been observed in captive shrews such as *Sorex araneus*, little is known about this habit and its functions.

FOOD NICHE OVERLAP AND MODES OF ECOLOGICAL SEPARATION: THE PROBLEM OF COMPETITION

Being common, ubiquitous, highly active and voracious small mammals, shrews might be expected to come into competition with each other for food. This problem is partly overcome by geographical separation and differences in habitat preferences between species, but many communi-

*Figure 5.9 Common shrew (*Sorex araneus*) indulging in refection*

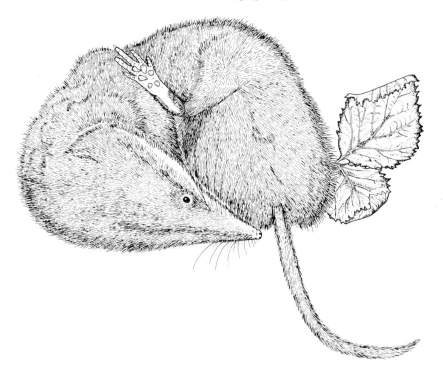

ties of shrews comprise more than one species coexisting in the same area. Commonly, two or three species coexist, as in the case of *Sorex araneus*, *S. minutus* and *Neomys fodiens* over much of Britain and Europe, and *S. cinereus*, *S. fumeus* and *Blarina brevicauda* in eastern North America. However, communities of five or six species are not unusual. For example the European species just mentioned may be joined by *Crocidura russula* and *C. leucodon*, as found by Yalden, Morris and Harper (1973) in an area of grassland in north-western France, and the American species may be accompanied by *S. hoyi*, *S. dispar* and *S. palustris*. Up to ten different species have been caught in primary forest and swamp in Zaire (Dieterlen and Heim de Balsac, 1979). Thus, many species of shrews coexist in the same habitat. How do such species avoid competition with each other?

Studies of communities of lizards (e.g. Pianka, 1973) and small desert rodents (e.g. Brown, 1975) indicate that differences in diet and foraging mode are important means of reducing competition between species which are otherwise similar in habits. Could this be so in shrews?

Shrews have such diverse diets that it is often difficult to detect obvious differences between species. Dietary overlap in terms of the prey types shared is often very considerable. For example dietary overlap between *Sorex cinereus* and *S. longirostris* was calculated at 77 per cent (French, 1984), and between *S. araneus* and *S. minutus* at 84 per cent (Church-field, 1984b). Even the water shrew (*Neomys fodiens*) I found, had a dietary overlap with the common shrew (*Sorex araneus*) of 79 per cent, despite its aquatic habits. However, studies of common, pygmy and water shrews showed that overlap was reduced by each species taking different proportions of major prey types. The overlap between common and pygmy shrews (the most similar species) was reduced from 84 per cent to 58 per cent by this means (Churchfield, 1984b, in press).

Overlap between different species may also be achieved, as I have suggested previously, by the adoption of different foraging modes. For example, investigations by Terry (1981) of four coexisting insectivores revealed that the shrew mole *Neurotrichus gibbsi* burrowed deeply in the soil; the shrew *Sorex trowbridgii* also burrowed but was commonly found on the ground surface foraging amongst litter and moss; *S. monticolus* was more restricted to the litter layer on the forest floor and *S. vagrans* did not burrow but was found more in open areas where numbers of *S. trowbridgii* were low. Hence, interspecific competition may be reduced by even quite small and subtle differences in habitat use for foraging.

Similar differences in habitat use have been found in other shrew communities, particularly with respect to hypogeal and epigeal habits, as Figure 5.6 shows. For example, although habitat segregation was not exclusive, I confirmed the observations of Michielsen (1966) and Pernetta (1976) that *S. minutus* is primarily epigeal, taking about 60 per cent of its prey from the ground surface and only 10 per cent from the subsurface soil while *S. araneus* augments surface feeding (30 per cent) by taking about 40 per cent of its prey from the soil.

Ecological separation on the basis of prey sizes exploited has been demonstrated for lizard communities (Pianka, 1973) and for small rodents (Brown, 1975) and there is some evidence of this in shrew communities, but the relationship is far from clear-cut since all species exploit prey of many sizes. For example, *Sorex araneus* and *S. minutus* often coexist with *Neomys fodiens*, and there is a clear difference in size between the three species (see Figure 5.10). Despite this, each shrew does not restrict its diet to a particular size range of prey, and the tiny pygmy shrew can take quite sizeable prey while the later water shrew eats many small prey (see Figure 5.11). Nevertheless, the diminutive *S. minutus* does tend to take a greater proportion of smaller prey than the larger shrews, reducing overlap with its closest competitor (*S. araneus*) to about 57 per cent.

The amount of dietary overlap also appears to differ according to the number of species present in the community. Overlap was greatest when only two species were present but increased when three species were present (Churchfield, in press). Thus shrew communities appear to conform to ecological theories about niches: niches are squeezed with more species present but expand when the number of competing species is reduced.

Of course, another factor which will affect dietary overlap and competition between species is the availability of food. If food is plentiful, then there is enough for all and competition ceases to be an important issue. For instance, overlap between shrews in terms of prey size appears to be a reflection of the prey sizes available. I found that although the greatest biomass occurred as large prey (> 20mm in body length), small prey (3–10mm) represented the most numerically abundant and also the most diverse food source. So shrews would be more likely to encounter small

Figure 5.10 Differences in body size between pygmy, common and water shrews (Sorex minutus, S. araneus *and* Neomys fodiens)

Body weight (g)	Body length (mm)	Tail length (mm)	
2.3–5.5	40–55	32–46	 Pygmy shrew
5.0–13.0	48–71	48–80	 Common shrew
9.0–16.0	61–72	45–77	 Water shrew

rather than large prey. If they are unselective in the face of competition, as the studies of Barnard and Brown (1985a) suggested, then all species of shrew would be expected to eat prey of all sizes, including small prey. Beetles are a good example: not only are they important prey items in terms of abundance and biomass but they also have the best overall food value in terms of high energy and low water content. So it is not surprising that they form a major component of many shrews' diets.

THE EFFECTS OF SHREWS ON PREY POPULATIONS

Studies reveal that shrews do have a measurable effect, even if not a controlling influence, on invertebrate communities. Studies by Platt and Blakley (1973) suggested that changes in levels of predation by the masked shrew (*Sorex cinereus*) produced certain differences in invertebrate prey populations. They took advantage of natural differences in the population densities of masked shrews in various parts of an area of prairie grassland to examine the relationship between predator and prey. They monitored shrew populations by live-trapping, and invertebrate populations by pitfall trapping.

They found that increases in the population density of shrews were associated with a decrease in the total numbers, but not the biomass, of invertebrate prey. There was also a general decrease in species

*Figure 5.11 Dietary composition of common, pygmy and water shrews (*Sorex araneus, S. minutus *and* Neomys fodiens*) with respect to the prey sizes taken (after Churchfield, in press)*

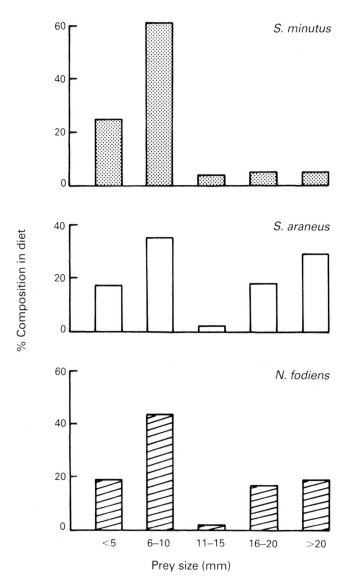

dominance but an increase in species richness and diversity amongst the invertebrate population which accompanied higher levels of predation by shrews. For example, when shrew density was low, large crickets were dominant components of the invertebrate fauna in terms of numerical abundance, and comprised a major proportion of all invertebrates caught in pitfall traps. But their dominance declined as shrew numbers increased. The reduction in their numbers, presumably by shrew predation, was compensated by an increase in the numbers of other, often smaller, types of invertebrate. From this they concluded that shrews do not affect the overall biomass of the prey they feed on so much as alter the numbers and composition of the prey population.

Increased shrew density was also associated with a decline in numbers of carnivorous invertebrates collected in pitfall traps, which highlights an important aspect of the feeding habits of shrews and the effect they have on invertebrate population structure. Shrews eat large numbers of carnivorous, ground-dwelling beetles (particularly Carabidae and Staphylinidae), spiders and centipedes. In so doing, they may well affect the guild structure of invertebrate populations, allowing herbivorous and detritivorous invertebrates to increase in number as their predators are reduced. With their high metabolic rates, large daily intake of food relative to body size, great activity and mobility, shrews may be keystone predators in a variety of terrestrial ecosystems.

The studies by Platt and Blakley were based on observations of wild populations of shrews, and looked at trends in invertebrate numbers and biomass which were coincident with the occurrence of shrew populations of different sizes. The only way of judging the true impact of shrews as predators is to perform controlled experiments where shrews and their prey can be manipulated in the field. Such an approach is being adopted by myself, Hollier and Brown with the use of small field enclosures which allow free entry to invertebrates but exclude shrews. Invertebrate numbers and diversity can then be compared in control plots where shrews have free access and adjacent plots where they are excluded. Results of these studies are revealing that, although shrews do not cause a massive depletion of prey populations in invertebrate-rich grasslands, they do significantly reduce numbers of the larger prey types such as spiders, myriapods, woodlice, slugs and snails and insect larvae.

Where several species of shrew coexist in a community, feeding on a wide spectrum of prey types and sizes, the effects may be greater. Churchfield and Brown (1987) estimated that predation by a small mammal community on invertebrates in grassland reach 6,800 prey items per hectare per day, due primarily to the activities of *Sorex araneus* and *S. minutus*. They concluded that the combined predatory impact on insect populations alone averaged 0.01 per cent per day. Although this may not sound much, it amounted to the equivalent of shrews clearing $1.1m^2$ of insects every day.

6 Energetics of shrews

Being amongst the smallest mammals, shrews are of considerable interest in the way in which they balance their energy budgets and overcome a wide variety of environmental conditions. Their small size imposes a number of constraints upon their physiological design, and yet shrews have evolved to cope with these by various intriguing means. It is the strategies that they have adopted to permit them to survive and flourish under adverse climatic conditions which will be the main substance of this chapter.

BODY SIZE AND THE METABOLIC RATE OF SHREWS

The size of an animal has very important effects on its physiology, including its energy expenditure as reflected in its metabolic rate. There is a general relationship between overall metabolic rate and body size in terms of mass. There is also a relationship between the metabolic rate per unit mass of tissue (the so-called mass-specific rate) and the whole body mass, as is shown in Figure 6.1. Overall metabolic rate obviously rises with increasing body mass, whereas the mass-specific metabolic rate decreases with an increase in body mass. These relationships are clearly demonstrated by the examples of metabolic rates in mammals of different sizes in Table 6.1.

Smaller mammals have more mitochondria per unit volume of tissue than larger mammals, which is one reason why the metabolic rate of a gram of tissue of a small animal is higher than that of a large one. The difference in mass-specific metabolic rate between large and small mammals helps to compensate for the higher rate of heat loss from a small mammal due to its relatively large surface-area-to-volume ratio, and large surface area per unit mass.

It is possible to predict, using mathematical equations, the metabolic rate of an animal of a known size. Using examples of many different mammals of varying body sizes, from mice to elephants, Kleiber (1961) showed that the basal metabolic rate in mammals is proportional to total body mass raised to the power 0.75 and to mass-specific-rate raised to the

Figure 6.1 The general relationship between metabolic rate and body mass in mammals

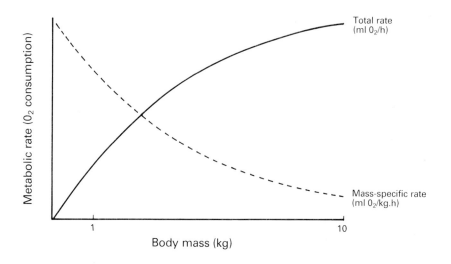

Table 6.1 Metabolic rates in mammals of different sizes

	Body mass (g)	Total O$_2$ consumption (ml/h)	O$_2$ consumption per g (ml/g.h)
Shrew (*Sorex*)	4.8	35.5	7.40
Harvest mouse	9.0	22.5	1.50
Kangaroo mouse	15.2	27.3	1.80
Rat	290	250	0.87
Cat	2,500	1,700	0.68
Horse	650,000	71,100	0.11

power −0.25. However, some mammals do not conform to this simple scaling relationship based on body size, but have considerably higher mass-specific metabolic rates than is predicted by the Kleiber curve. Shrews provide an interesting example of this. If we look at Table 6.1, it can be seen that the metabolic rate of the shrew falls into a category of its own. As a result of the apparent peculiarities of their energy relations, their high metabolic rates and extremely small body sizes, shrews have been the subject of great interest to those studying energetics.

The very small size of shrews has important consequences for their rate of energy expenditure, their ability to store reserves of energy in the form of body fat and, therefore, on their survival abilities. As I have said, shrews, being small, have a large surface-area-to-volume ratio. Heat loss is rapid and this has to be compensated for by increased heat production by way of a high metabolic rate. A we have already seen, fat storage in most shrews is minimal, and of uncertain survival value. Because of this they are unable to hibernate in adverse conditions, such as during the winter, and must remain active in their search for food every day throughout the year. The energy budget of small homeothermic animals like shrews is critically balanced, with only a fine line between starvation and survival. Unless they have evolved physiological and/or behavioural adaptations which help reduce their energy needs, especially during adverse conditions, they are dependent on a regular and frequent intake of energy to compensate for their high rate of energy expenditure.

Despite these problems, and as we saw in Chapter 1, shrews are highly successful and ubiquitous small mammals. They have colonised all kinds of terrestrial and semi-terrestrial habitats at various altitudes, and they have a wide geographical distribution extending from tropical regions to the cold tundra of the far north. How do they manage to survive under such conditions?

Estimates of the metabolic rates of shrews have been the subject of considerable controversy for many years. This is due to the immense difficulties of making accurate measurements on such small, highly active and excitable animals under captive and confined conditions. The most usual technique for measuring basal metabolic rate is to confine the animal in an airtight vessel and monitor the amount of oxygen consumed by it in a given period and at a known ambient temperature. In order for the basal metabolic rate (BMR) to be estimated and for results of different trials to be comparable, the animal should not be stressed or highly active, since this will increase energy expenditure, and hence metabolic rate. For shrews, it has proved very difficult to obtain a resting or basal rate under experimental conditions, and widely differing results have been obtained, leading to a debate about whether they do or do not have peculiarly high metabolic rates for their size.

More recently, techniques have been refined, and more accurate results obtained. The results of such studies show that body size does have an important effect on daily energy expenditure (or average daily metabolic rate, ADMR) in shrews. This can be seen in Figure 6.2 where body mass is plotted against the absolute energy expenditure of captive individuals under constant conditions. To provide a comparison with other mammals of similar size, the average daily energy expenditure at 20°C for small rodents is included, and is represented by the line R. As can be seen, the majority of shrews studied lie above this line. Of those shown on the graph, only *Blarina brevicauda*, *Crocidura russula* and *Suncus etruscus* have relatively low metabolic rates which are comparable to rodents of similar size.

The situation regarding body size and metabolic rate is complicated by those species, particularly in northern temperate regions, which undergo seasonal changes in body weight, since this directly affects absolute energy expenditure, as can be seen in the examples of *Neomys fodiens* and *Sorex minutus* in Figure 6.2. *Sorex araneus* may increase its absolute energy expenditure by some 25 per cent from a sub-adult to an adult on the basis of increasing body weight alone.

Some caution is required in the interpretation of these results, however, because estimates of energy requirements and expenditure may well be very different in wild shrews compared with the captive ones which feature in most of the experiments reported so far. Captive shrews kept under constant laboratory conditions are not subject to the vagaries

Figure 6.2 Daily energy expenditure of shrews of different body sizes. The dotted line R is the average value for rodents. Open symbols represent Crocidurinae (Cr: Crocidura russula; Se: Suncus etruscus). Closed symbols represent Soricinae (Bb: Blarina brevicauda; Na: Neomys anomalus; Nf: N. fodiens; Sa: Sorex araneus; Sc: S. coronatus; Sca: S. caecutiens; Si: S. isodon; Sm: S. minutus; Smi: S. minutissimus). Different measurements for the same species are joined by solid lines (after Genoud, 1988)

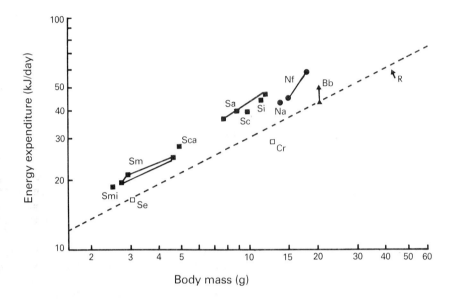

of climatic conditions and food availability as are wild ones. Certainly, the behaviour and activity of wild shrews are very different from those of laboratory-kept ones. The thermoregulatory costs of shrews in winter may boost energy expenditure to several times the basal rates, even when they are only resting in their nests.

Nevertheless, the results show that shrews fall into two quite distinct categories with respect to their metabolic rates. As we have already seen, shrews are divided into two major taxonomic groups or subfamilies, the Soricinae and the Crocidurinae (see Chapter 1). The former are predominantly northern in distribution while the latter are tropical in origin. These two categories also reflect their physiological characteristics, for the Soricinae typically have very high metabolic rates which are considerably greater than those predicted from the Kleiber curve. Members of the Crocidurinae, on the other hand, have lower metabolic rates than soricines of a similar size, and their rates are similar to, or only slightly higher, than those expected on the basis of the Kleiber relation. This can be seen in Figure 6.3.

In fact, shrews exhibit a wide range of basal metabolic rates (see Figure 6.3 and Table 6.2). For example, metabolic rates of *Blarina brevicauda*, *Neomys fodiens* and *Sorex vagrans* have been found to be greater than 150 per cent of the expected value. While some of the very high rates recorded may not relate to resting condition, basal rates over 200 per cent of the expected value have been found in some species, namely

Figure 6.3 The relationship between basal metabolic rate and body mass in shrews. Line K is the Kleiber curve and line B is the boundary curve for continuous maintenance of endothermy (see text). Crocidurinae: Cl: Crocidura leucodon; Co: C. occidentalis; Cs: C. suaveolens; Smu: Suncus murinus. Soricinae: Nc: Notiosorex crawfordi; Ss: Sorex sinuosus; Sv: S. vagrans. All other symbols as for Figure 6.2. Large symbols represent reliable estimates of BMR; small symbols represent values possibly obtained under non-basal conditions (after Genoud, 1988)

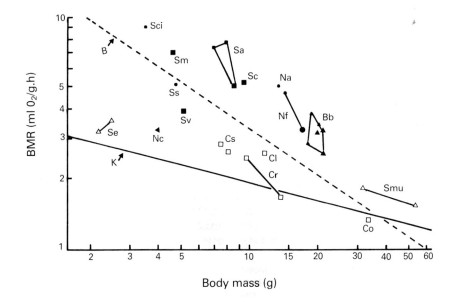

S. araneus, S. coronatus and *S. minutus*, under carefully controlled conditions where stress and activity were minimal. Not all Soricinae have such high rates: the desert shrew *Notiosorex crawfordi*, has a metabolic rate only slightly greater than expected from the Kleiber curve and is thus more characteristic of the Crocidurinae. This is undoubtedly related to its way of life and the habitat (hot desert) in which it lives, which resembles that of the crocidurine shrews.

One further correlate of body size and metabolic rate is the life-span of an animal. It has been demonstrated that the mean length of life is positively correlated with body size in shrews (Hutterer, 1977). Moreover, the Crocidurinae have longer lives on average than the Soricinae of equivalent size classes, which is probably related to their lower metabolic rates.

MAINTAINING BODY TEMPERATURE UNDER ADVERSE CONDITIONS

Shrews maintain body temperatures which are similar to those of other mammals but, again, body temperature does differ between the two major groups. The Soricinae have an average temperature of 38.5°C, slightly above that of other mammals. The Crocidurinae have a lower

115

Table 6.2 Metabolic rates and body temperatures in different species of shrew (modified from Genoud, 1988)

	Body weight (g)	BMR ml O_2/g.h	Body temperature °C
Soricinae			
Sorex minutus	4.6	7.04	—
S. minutus	—	—	38.1
S. cinereus	3.5	9.0*	38.8
S. vagrans	5.2	3.79	—
S. araneus	7.8	7.74*	—
S. araneus	8.7	—	38.3
S. coronatus	9.4	5.20	—
Notiosorex crawfordi	4.0	3.27	37.6
Neomys fodiens	17.1	3.22	37.3
Cryptotis parva	3.6	—	35.0
Blarina brevicauda	20.3	3.18	37.5
Crocidurinae			
Suncus etruscus	2.5	3.60	—
S. murinus	52.0	1.66	—
Crocidura suaveolens	7.5	2.81	34.8
C. russula	9.6	2.45	34.5
C. leucodon	11.7	2.55	34.2
C. occidentalis	33.2	1.34	—

* Possibly not basal rate

body temperature, averaging 35.5°C (Frey, 1979) (see Table 6.2). But this strict difference between the two groups has exceptions, for the lowest temperature is maintained by *Cryptotis parva*, a member of the Soricinae, at 35°C. The Soricinae also regulate their body temperatures more strictly than the Crocidurinae under normal conditions, for example within ±2°C in members of the genus *Sorex*, *Neomys* and *Blarina*, but within ±5°C in *Crocidura* and *Suncus*.

Maintaining a constant, and high, body temperature under changing ambient conditions has important implications for heat production and metabolic rate, and climatic conditions have had an important influence on the evolution of the two strategies adopted by shrews. This is particularly so for shrews in northern temperate regions where, in order to maintain their body temperature and not suffer from hypothermia, they must increase heat production. To do this they must increase their metabolic rate to fuel the extra heat production. Studies on the American short-tailed shrew, *Blarina brevicauda*, have shown that there is a seasonal variation in basal or resting metabolic rate and the rate of thermogenesis (heat production), both showing a maximum in the winter (Merritt, 1986). There was a 38 per cent in metabolic rate and a 54 per cent increase in the capacity for non-shivering thermogenesis (NST) in January compared with August. NST is a rapid form of non-shivering heat production in mammals which occurs particularly at sites of brown adipose tissue (BAT) accumulation. This brown fat produces a large amount of heat as it is metabolised. Accompanying this was a seasonal variation in the amount of brown fat stored by *Blarina*, with the interscapular mass of brown fat being considerably greater in winter than in summer.

Brown fat also undergoes seasonal variation in *Sorex araneus*, reaching a peak in young shrews in autumn and winter (Churchfield, 1981) or in spring when these shrews are reaching sexual maturity (Pasanen and Hyvarinen, 1970). However, the amount of fat stored by shrews is small compared with many other mammals, and its survival value is doubtful. Investigations of seasonal metabolic rates of other shrew species have produced some equivocal results. The average daily metabolic rate of *Sorex araneus* is reported to be somewhat lower in autumn and winter than in spring and summer (Gebczynski, 1965).

Heat loss can be reduced by increasing the insulative properties of fur and reducing heat conductance through the skin. Shrews undergo seasonal moults and those exposed to the low ambient temperatures of a northern winter are aided by a pelage which is longer, thicker and darker than the summer pelage. The heat loss of sub-adult *S. araneus* in autumn and winter is reported to be only marginally higher (about 6 per cent) than in young shrews in summer (Gebczynski, 1965). In *Blarina brevicauda* heat loss is lower in winter compared with summer (Randolph, 1973).

Shrews in tropical and sub-tropical regions are also subjected to extreme ambient temperatures, but it is the capacity to withstand high temperatures which is the key to their survival. A high metabolic rate in a small mammal inhabiting a hot climate has major disadvantages. It results in high water loss, overheating and, ultimately, death. It can even give rise to thermoregulatory problems in quite moderate environments. For example, *Sorex cinereus* cannot survive at temperatures greater than 24°C when sitting in its nest, and the soricine shrews found in the warm Mediterranean region occur mainly in wetter habitats where heavy water losses can be most easily compensated for.

Many of the *Crocidura* species, originating in hot, tropical environments, are known to be better at coping with higher ambient temperatures than species of *Sorex*, *Blarina* and *Neomys*. So, too, is the American desert shrew *Notiosorex crawfordi*. These heat-adapted shrews have a thinner pelage and higher rate of conductance of heat through the skin, plus a relatively low metabolic rate which reduces heat production. They may also employ daily torpor to save energy and reduce water loss during the hottest parts of the day, remaining in the shade of rocks, shrub roots and subterranean burrows. They can then restrict their foraging activities to the cooler periods.

An interesting situation arises with those shrews with a semi-aquatic habit such as *Neomys fodiens* and *Sorex palustris* which swim and dive underwater for food. This presents particular energetic problems in a cool climate because heat loss in cold water is greater than in air. Studies have suggested that the body temperature of these two shrews is influenced by water temperature and the period of immersion, and that body temperature decreases significantly with decreasing water temperature. For example, the coefficient of heat transfer of *S. palustris* in water was found by Calder (1969) to average 4.6 times that in air. This put an extra burden on maintaining homeostasis in the body, and the result was a reduction in body temperature of about 2.8°C per minute in forced dives in water of 10–12°C. Foraging in water, particularly in winter, could be a major energetic problem under these conditions.

In order to make the most of this apparently energetically costly foraging strategy, these shrews should only use it when aquatic prey are abundant and readily accessible and terrestrial prey are in short supply. It seems that neither *N. fodiens* nor *S. palustris* restrict themselves entirely to aquatic prey, and that they both mostly frequent quite shallow waters which are rich in invertebrates. However, recent studies by Vogel (1990) have demonstrated that a key factor in the maintenance of body temperature in these aquatic shrews is their general state of health and, particularly, the condition of the fur. He found that the body temperature of healthy *N. fodiens* changed according to circumstances, even before it entered the water. While at rest, shrews had a body temperature of 37°C. This rose to 37.5°C when they were active in a terrestrial habitat, and to 39.4°C while in a social confrontation with other shrews. When in water, a shrew with wet fur underwent a decrease in body temperature of 1.1°C per minute during forced swimming. Shrews kept under good conditions of captivity where they could maintain the high quality of their pelage were able to keep their fur dry during swimming and could maintain their body temperature of about 37°C in a variety of test situations. For example, in water at 2.6°C, the mean body temperature after six minutes of forced swimming and diving was 37.4°C, comparable to that during terrestrial activity. Vogel found that the fur remained dry, even on the surface.

In the wild, and under good conditions of captivity, the pelage of these shrews has unique hydrophobic properties which effectively repel the water and insulate the body most efficiently. In captivity, the fur of shrews often goes out of condition, probably due to a change in diet and the substrate provided. The fur loses its dense, clean and glossy look and becomes greasy and matted, quickly losing its waterproof properties. It is likely that many experimental studies of temperature loss in water shrews were undertaken on shrews with poor coat condition, which had led to some misleading results. Contrary to expectation, then, aquatic foraging may not seem to be such an energetic drain on shrews, even in winter. The shrews themselves confirm this, since they forage in water throughout the year, in all seasons and weathers.

It is interesting to note that these water shrews are not very well adapted for a semi-aquatic life of swimming and diving. They have only rudimentary anatomical adaptations of which the thick, water-repellent pelage is probably the most useful to them. Their large size relative to other shrews gives them a greater thermal inertia and a lower rate of heat loss, but they have no special physiological adaptations to assist diving. They are unable to remain submerged for longer than a few seconds but, since heat loss could be a major problem with increasing length of immersion, this may be their way of avoiding hypothermia.

Very little is known about the lifestyle of the other aquatic shrews such as the Tibetan web-footed water shrew of the genus *Nectogale* which is, anatomically at least, rather better adapted for swimming than *Neomys* and *Sorex*.

MAKING THE BEST USE OF
THE ENERGY AVAILABLE

As we have already seen, shrews generally have a characteristically high daily food consumption which is commensurate with their high metabolic rates. Examples of food consumption by shrews of different body sizes are shown in Figures 5.3 and 5.4, and the subject has been discussed in Chapter 5. As with the results of oxygen consumption as a measure of metabolic rate, estimates of daily food intake show that as body mass increases so does absolute food consumption. But food intake as a percentage of the body mass decreases with increasing body size. In fact, food consumption can be used as a measure of metabolic rate and daily energy requirements, but the results tend to be more variable than those obtained from using respirometry techniques.

Shrews make good use of the energy available to them in their food. They generally have high assimilation efficiencies (the amount of food energy which is digested and available for metabolic processes and growth), despite the large amount of indigestible chitin ingested with many prey. They have assimilation efficiencies similar to granivorous rodents but higher than herbivorous rodents of equivalent size. However, there is a tendency for assimilation efficiency to decrease with increasing body size. In shrews maintained on a diet of blowfly larvae, assimilation efficiency ranged from 70 to 95 per cent, but that of *Neomys fodiens*, for example, was significantly lower than that of *Sorex araneus* and *S. minutus*. The white-toothed shrews (*Crocidura*) have rather lower efficiencies than the other species, which is probably related to their lower metabolic rates. The actual efficiency varies, depending upon the kind of prey eaten. Earthworms, insect larvae and other soft-bodied prey are likely to be digested more efficiently than are those with hard, chitinous exoskeletons. However, with large prey the most chitinous and least profitable parts, such as the legs of spiders and beetles, are often not eaten.

There is no evidence that shrews attempt to restrict their diets to the most energetically profitable prey. An examination of the diets of wild shrews suggests that there is little selection for prey which are easy to digest or have low chitin contents. For most shrews the diet is a highly varied one, although certain prey may feature prominantly as food items (see Figure 5.1). The diets appear to be more closely correlated with the availability of prey than with their relative food values. However, invertebrates show only small differences in food value in terms of energy content. Figure 6.4 shows the energy and water contents of a range of major prey types of shrews. Only Isopoda (woodlice) are notable for their very low energy value, and yet they are frequently eaten by shrews, and for species such as pygmy shrews (*Sorex minutus*) they are a dominant prey item. Water content is more variable and, by making up a lot of the bulk of many invertebrates it reduces their total food value. Earthworms, slugs and snails are very high while beetles are low in water content. Overall, beetles have the best trade-off between energy and water contents, and they are frequently very abundant. It is not surprising, then, that they feature as major dietary items for many shrews.

It is interesting that shrews taken into captivity and placed on uniform

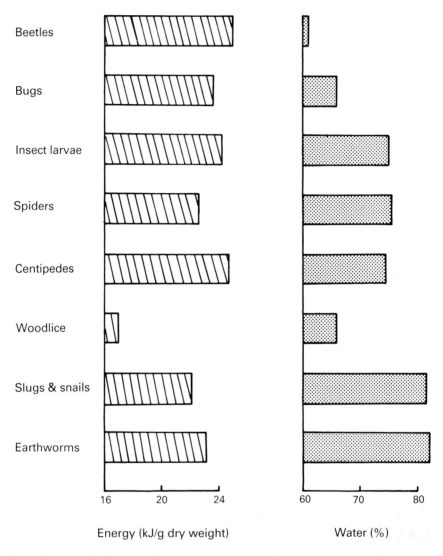

Figure 6.4 Energy values and water contents of different invertebrate prey commonly eaten by shrews

Energy (kJ/g dry weight)   Water (%)

diets of invertebrates such as mealworms or fly pupae lose condition very quickly. So a varied diet is obviously necessary in order to obtain the nutrients they need. This must also be an important factor affecting their foraging habits. Their high metabolic rate probably affects their foraging strategies, too, and makes them relatively unselective as far as prey items go. They can only satisfy their energy needs by being opportunists and taking whatever prey they find. They cannot afford to reject potential prey items.

The availability of food resources may also have influenced the development of different metabolic rates in shrews. Possession of a high metabolic rate necessitates a regular and frequent food intake, and a high one. This is a major disadvantage in environments where the food supply is unpredictable and may be subject to great fluctuation. In hot, dry tropical and sub-tropical climates there may be flushes in vegetation and invertebrate numbers coincident with the rainy season, but food may be in short supply for long periods in between. These conditions are best survived by having a relatively low metabolic rate together with an ability to enter torpor. In temperate climates food resources may fluctuate seasonally, but are relatively more predictable in supply throughout the year. Shrews need a high metabolic rate to survive a cold climate anyway, so it is not possible to lower it, even if food is scarce.

Although a decrease in ambient temperature might be expected to increase energy requirements, this does not necessarily result in an increase in food consumption by shrews. Table 6.3 shows the food

Table 6.3 Food consumption and assimilation by common shrews (*Sorex araneus*) at winter and summer temperatures (after Churchfield, 1982b)

	Summer (20°C)	Winter (1–9°C)
Mean body mass (g)	8.1	8.2
Mean food intake (g)	8.3	6.4
Intake as % of body mass	97.1	79.2
Mean assimilation efficiency (%)	85.5	84.4

consumption and assimilation efficiencies of common shrews (*Sorex araneus*) maintained in captivity under summer and winter conditions of temperature and daylength. Contrary to expectation there was a significant decrease in food consumption under winter conditions at 1–9°C compared with summer conditions at 20°C. There was no difference in assimilation efficiency at the two temperature regimes, and the body weights of the shrews varied less than 0.5g during the experiment, which lasted over 3–4 days. Homeothermic animals such as shrews, which can maintain constant body temperatures, are therefore able to keep their digestive processes going at full efficiency even under adverse environmental conditions.

The lowered food intake under winter conditions can be attributed to a change in the activity pattern of the shrews. At cold temperatures they spent much less time outside their nests than they did at warm temperatures. In this way they lowered their energy requirements by reducing heat loss as well as locomotory energy costs. The assimiliation efficiency of *Blarina brevicauda* is also unaffected by changes in ambient temperature (Randolph, 1973).

The mechanism of losing weight and decreasing body size in winter may help shrews to compensate for the increased energy use via heat production at lower ambient temperatures by lowering their absolute food requirements. In this context it is interesting to note that *Sorex araneus* from poor habitats (such as spruce forest), where population density is low, have been found to be smaller than shrews from more productive habitats (like grasslands) where population density is higher (Heikura, 1984).

Is the winter size decrease of shrews, so characteristic of *Sorex* species, really a strategy to reduce absolute food requirements in winter? What difference would a loss in weight make to the food requirements of a shrew? From a knowledge of the rates of food consumption of shrews, I calculate that a loss in weight of some 30 per cent could decrease the food requirements of a common shrew (*S. araneus*) by about 20 maggot-sized prey (each of 10mm in length) per 24 hours. In turn, this would reduce the necessary foraging time by about five hours per day. This supports the idea that the winter weight loss has evolved as an energy-saving strategy. However, this is only speculation and, indeed, it may not be a strategy in all species. For example, Merritt (1986) reports that *Blarina brevicauda* does not undergo a winter size decrease, despite inhabiting areas with a cold temperate climate. Instead, this relatively large shrew shows a steady increase in body mass over the winter, suggesting that the availability of prey must be sufficient not only for daily maintenance costs but also to permit growth. So *Blarina* does not seem to be food limited in winter.

It is also interesting to note that *Sorex araneus* kept in outdoor enclosures tends to lose weight in winter, even when a plentiful supply of food is provided. So perhaps the weight-loss strategy is independent of food supply in *Sorex*.

Nevertheless, northern representatives of shrew species are smaller than more southerly ones, the coldest regions being inhabited by the smallest species of *Sorex*, and this is sometimes seen as an adaptation to cope with a low food supply. Such a strategy has been observed in moles which are also non-hibernators. Selection for small body size has been found in mole populations after an extremely hard winter (Stein, 1950), the implication being that only the smaller individuals survive.

TORPID AND NON-TORPID SHREWS

The ability to enter torpor in mammals is related to body mass and basal metabolic rate. Shrews are unable to hibernate, even though some mammals of a similar size can do so. Unlike many bats, which are capable of reducing their metabolic rates and body temperatures drastically to allow them to hibernate for several months, shrews are not adapted to do this. The metabolic rates of shrews may vary seasonally, but only within a fairly small range for each individual. Similarly, body temperatures can only be maintained within a narrow range compared with those of bats, for example. Fat reserves are also not sufficient to support a shrew during prolonged periods of inactivity. The failure of shrews to evolve hibernation as a survival strategy may reflect a lack of selective pressure in respect of seasonal food supply. Aerial insects (such as flies and moths) on which insectivorous bats feed are in very short supply in winter, whereas ground-dwelling invertebrates on which shrews subsist are available all year round.

Nevertheless, some shrews are capable of undergoing short periods of torpor lasting several hours. Several members of the Crocidurinae have shown the ability to enter shallow daily torpor, including *Crocidura russula*, *C. suaveolens*, *Diplomesodon pulchellum* and even the tiny

Suncus etruscus, when subjected to food shortage. Members of the Soricinae, on the other hand, are generally not able to do this and, again, this is related to their higher metabolic rate.

On the basis of the relationship of torpor with body mass and BMR, McNab (1983) devised an allometric equation describing a minimum boundary curve which separated species that could enter torpor from those that could not (see Figure 6.3). The minimal boundary curve represents the limit for continuous maintenance of endothermic temperature regulation. As we see from Figure 6.3, the Soricinae, with their high metabolic rates, mostly lie above this curve and are generally unable to alter their body temperatures and enter torpor. The Crocidurinae mostly lie below the curve; they are able to lower their body temperature by a few degrees, slow down their bodily processes a little and enter torpor for a few hours. An exception is *Notiosorex crawfordi* which, although it is a soricine, behaves like a crocidurine with a basal metabolic rate below the minimal boundary curve and the ability to enter torpor. We would predict from their position just beneath the curve that *Sorex vagrans* and *S. sinuosus*, also soricines, should be able to enter torpor. In fact, *S. sinuosus* has been found to exhibit torpor-like behaviour. *Suncus murinus* is another oddity for, although it is a large crocidurine, it lies just above the line and apparently cannot enter torpor.

Little is known about the actual energy-saving to shrews if they can undergo torpor. But Genoud (1988) has estimated that torpor results in a saving of less than 15 per cent of the daily energy expenses in captive *Crocidura russula* and *Suncus etruscus*, so it seems to be just a short-term strategy to allow these shrews to survive an immediate energy crisis such as a shortage of food.

ACTIVITY AND REST

The activity of an animal also has a direct effect on its rate of metabolism. For example, the metabolic rate of *Blarina brevicauda* during normal exploratory activity is approximately twice that of its sleeping rate (Randolph, 1973). The energy expended through activity is related to an animal's body mass as well as to the rate of activity. The locomotory cost involved, even during the normal activity of foraging and exploring (without excessive exertion) is very high in a small compared to a large animal. Even amongst shrews of different body sizes this is quite apparent, *Suncus etruscus* being a particularly good example (see Table 6.4).

So being small has another penalty which has to be compensated for by increased energy consumption. But in order to forage efficiently, a shrew must cover as much ground as possible in its search for food. Prey are generally not uniformly distributed, but tend to be clumped in areas which meet their particular environmental requirements, and clumps have to be located. A shrew must also move quickly around its home range to find sufficient prey to meet its daily needs. A foraging shrew always runs rapidly through the undergrowth, pausing momentarily to probe with its snout for prey and then moving quickly on.

Most shrews must feed regularly every two or three hours or they will

Table 6.4 Energy costs of locomotion in shrews of different sizes (after Genoud, 1988)

	Body weight (g)	Locomotory energy cost (J/g.h)
Suncus etruscus	2.2	202
Sorex coronatus	9.5	105
Crocidura russula	11.0	97
Blarina brevicauda	20.0	54

die of starvation, so they must remain active both day and night. Periods of activity are spent bustling through the undergrowth or along subterranean tunnels as they search for prey and explore their home ranges. Bouts of activity may last from about 30 minutes to two hours, and alternate with periods of rest in the nest. Prior to resting, *Sorex palustris* has been observed to walk round and round in tight circles within the nest (as dogs do before lying down), and it curls up in a tight ball to sleep. The usual sleeping posture of shrews is lying on the belly with the feet and tail underneath the body and the snout tucked in towards the chest. Sleeping periods last from twelve to 130 minutes in *Sorex palustris* (Sorensen, 1962). However, complete rest or sleep in shrews usually lasts for only a few minutes at a time and may be interrupted by nest repair, feeding on cached prey or grooming.

Grooming is performed mostly by scratching through the fur with the hind feet. But shrews do sometimes lick those parts within reach, particularly the feet and the anal area. Grooming helps to keep the fur in good condition and wild shrews, almost without exception, maintain a sleek and glossy appearance at all times. This is particularly important for water shrews which must retain a waterproof pelage. Periods of swimming are followed by rigorous shaking and scratching to remove excess water. Grooming may also help to dislodge some of the external parasites such as mites.

Bouts of intense activity during foraging and exploring may be interrupted suddenly by a short nap. Instead of returning to the nest, the shrew crouches down wherever it happens to be, tucking its head and snout into its chest, remains motionless for a few seconds, and then resumes it activities.

Although they are active by day and night, shrews have cycles of activity through the 24-hour period. Common shrews (*Sorex araneus*) seem to have about ten periods of almost continuous activity alternating with rest over the 24-hour period. Many shrews, including *S. araneus*, *S. vagrans*, *S. palustris* and *Neomys fodiens*, tend to be most active just before sunrise and after sunset, and least active in the afternoon. They also tend to be more active during darkness compared with light (see Figure 6.5), although *S. minutus* seems to be equally active by day and night. Shrews also show seasonal differences in their activity patterns. Common shrews (*S. araneus*) are more active and move greater distances in summer than in winter. In winter they spend more time in their nests or in subterranean tunnel systems than out on the ground surface.

Captive shrews, for example, were active for only 19 per cent of the 24-hour period during winter at cold ambient temperatures compared with 28 per cent at summer temperatures. Their rest periods were longer and activity periods shorter in winter, too (Churchfield, 1982b). This is confirmed by live-trapping studies on wild *S. araneus* and *Neomys fodiens* (see Figure 6.5). Similar observations have been made for *Blarina brevicauda*, the American short-tailed shrew, where around 30 per cent of the 24 hours is spent active in summer, but only 12 per cent in the coldest months of winter.

Figure 6.5 Diurnal-nocturnal activity of Neomys fodiens *(striped bars) and* Sorex araneus *(open bars) during different seasons of live-trapping (after Churchfield, 1984a)*

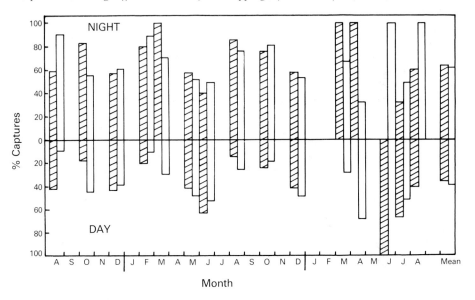

This reduction in activity in winter represents an important energy-saving strategy for shrews. They reduce heat loss and locomotory costs by decreasing the time exposed to cold ambient temperatures and spend longer in a well-insulated nest where *Blarina*, at least, may feed on cached food. Underground, temperature fluctuations are smaller and the environment is considerably warmer than on the surface in winter. *Blarina* and other northern shrews are able to remain active under the snow in winter, since the snow and leaf litter have an important moderating influence on temperature.

Shrews in tropical and sub-tropical regions have similar patterns of activity with alternating rest periods. *Myosorex cafer*, for example is active day and night but mostly during darkness. It may spend one or two hours at a time in its nest, just coming out briefly to forage and defecate. *Crocidura* species are also predominantly nocturnal and their overall levels of activity are generally lower than those for soricine shrews. These shrews are also rather slower and less agile in their movements and reactions than are *Sorex* species, which may be related to their lower metabolic rates.

The home range size of a mammal is mainly dependent on its energy requirements, the nature of its diet and its foraging behaviour. Again, home range varies according to body weight, with larger mammals generally maintaining bigger home ranges or territories than smaller mammals with similar feeding habits. However, in shrews there is considerable variation and home range size may not be clearly correlated with body size or energy requirements. For example, *Sorex coronatus* and *Crocidura russula* are sympatric in many areas of Europe and are similar in body size. Yet in winter they have very different home range sizes, even though they have similar energy expenses in the field (Genoud, 1985).

As we have seen, *Sorex minutus*, although very small, maintains a bigger home range than does the larger, sympatric *Sorex araneus*. Such variation is probably due to local and habitat differences in prey availability (and hence the energy value of the home range) and the mode of foraging. Thus, species of *Sorex* and *Blarina* tend to have home ranges which are similar in size to those of granivorous rodents which probably search for seeds in a similar fashion to the way shrews search for invertebrate prey. Seeds have a clumped distribution even greater than invertebrates, around the particular plant of their origin.

A large home range may have distinct advantages in foraging and energy gain. Since prey tend to be patchy or clumped in distribution, a large home range is likely to incorporate more food patches than a small one. If the home range is used in a systematic way, each patch is likely to be revisited and depleted of its prey less frequently in a large home range than in a small one, and so foraging success is improved. This would be of particular importance in areas where prey recruitment is slow. Maintaining a home range as a territory in which intruders are kept out may be energetically very costly unless regular patrolling can be combined with foraging.

OTHER WAYS OF SOLVING ENERGETIC PROBLEMS

Food Storage

Because of the fine balance between survival and death for shrews and their small energetic autonomy, most of them must constantly adjust the ratio of energy intake to energy expenditure in order to balance their energy budgets. They would be much less vulnerable if they were able to build up energy reserves in the form of body fat, but for wild shrews this does not seem to be an option. Given constant, warm conditions and plenty of food, captive shrews will quickly put on weight and build up large quantities of body fat. However, this never occurs in the wild, and less than 10 per cent of the body weight comprises fat in *Sorex araneus*, *S. minutus* and *Neomys fodiens* (Myrcha, 1969; Churchfield, 1981). Even though they do undergo a seasonal cycle in body fat content, this is of small magnitude and the amount stored would seem to have little value as an energy reserve during times of stress (see Chapter 2).

An alternative mechanism would be to make food hoards during times of plenty for use when conditions deteriorate, such as in winter.

Possession of a food store which was available when prey were scarce or difficult to locate would reduce the amount of energy expended in foraging for elusive prey, and the amount of heat energy lost while searching at low ambient temperatures. As we have seen, many captive shrews do readily indulge in food hoarding behaviour (Chapter 5), and this may be done in a very systematic way as soon as food is super-abundant or clumped in distribution so that it is easy to collect. Nevertheless, there is little evidence of wild shrews apart from *Blarina brevicauda* and *Neomys fodiens* making food hoards, and it is not clear whether their prey resources are ever distributed in such a concentrated way as to make food hoarding a viable strategy. Apart from certain species, such as *Blarina* and *Neomys*, which may feed on vertebrate prey, most shrews subsist on a diet of quite small invertebrates which, energetically speaking, are best eaten *in situ* rather than transported to a central store for future use. Moreover, the numbers of prey required by shrews, even on a daily basis, are prodigious and run into the hundreds. Building up a food store, as well as feeding itself as it went along, would require an immense amount of time and energy. Even if food hoarding does occur in the wild, a store of small invertebrates is unlikely to last more than a few days.

There are other problems involved with food stores, too. Firstly, the prey have to be incapacitated to prevent them from wandering off. This is usually done by mutilation and, in the case of earthworms stored by moles, decapitated or mutilated ones remain fresh for days, even weeks, in cool subterranean burrows. Smaller prey, such as beetles and woodlice do not seem to last as well, and shrews will not eat decaying invertebrates. There is also the problem of defending a food store against intruders trying to steal from it. If a shrew is out foraging for more prey it cannot be overseeing its store. At low population density and in situations of strict territoriality, a food store placed strategically at the centre of an individual's territory may avoid the attentions of unwelcome visitors of the same species. But where the home range is shared with other insectivorous species (not just shrews) or incursions into the home range by neighbours regularly occur, food hoards may not survive long.

The Use of Nests

The use of a well-insulated nest can greatly reduce energy expenditure from the body in the form of heat, as lovers of real down duvets well know. The resting metabolic rate of shrews can be reduced by up to 30 per cent by the use of a nest (Genoud, 1988). Shrews make quite large and complex nests with whatever materials are to hand. In the wild, nests found beneath fallen logs have been made from dried grass, shredded leaves, moss and even rabbit hair. In captivity, they will use many different materials including hay, straw, cotton wool, shredded cloth, tissue paper and wood shavings. Those kept in outdoor enclosures under near natural conditions will even bank up the edges of the nest with small stones collected from the surrounding soil.

Nest-making in captive shrews is a major activity. Nesting abilities seem to improve with age and practice. Young common shrews, for example, make only rudimentary nests with little structure, simply gathering a small and simple pile of materials together. Older shrews make more elaborate nests by gathering up larger quantities of materials and weaving dried grasses together to form a discrete spherical, dome-shaped or cup-shaped structure with several small entrances or exits. The nest of a pregnant female is even larger (up to 15cm in diameter) and more complex, and is constructed a day or two before she is due to give birth. Sorensen (1962) observed that nest construction by captive American water shrews (*Sorex palustris*), required endless trips to gather the necessary materials. He estimated that one nest, measuring 20cm in diameter, comprised over 100 pieces of shredded cloth. Nests were mostly about 8cm in diameter, but the largest was some 32cm. As it built the nest, the shrew placed itself at the centre of the gathered material and, by making a series of quick tight turns, it formed a small depression. From this central point it used its snout to stitch the materials of the walls together around it.

Video films of common and water shrews show that they spend a considerable time within the nest, not sleeping or resting, but making adjustments to the bedding material. The mouth is used to move pieces of grass around in the nest which are then pushed into place by the snout. The body is moved round and round the nest to open up the central area and flatten the sides. While captive shrews usually maintain a single nest, wild ones and those in large outdoor enclosures may have more than one distributed around their home range. Nests are periodically vacated and replacements constructed nearby.

Sharing a nest with a friend or relative reduces energy expenditure even further: in *Crocidura russula* and *Suncus etruscus*, nest-sharing results in a 20 per cent reduction in resting metabolic rate compared with that of a solitary nester (Genoud, 1988).

PUTTING ENERGY INTO GROWTH AND REPRODUCTION

Not only do shrews have to balance their energy budget on a daily basis, they also have to be able to set aside energy and resources for growth and reproduction. For females, the added cost of pregnancy and lactation is a particular burden which is increased with the size of the litter. It has been found for *Crocidura suaveolens*, at least, that the basal rate of metabolism increases during pregnancy and lactation (Genoud, 1988). Shrews can only grow and reproduce at times when enough resources are available to cover the additional costs of maintenance and production. So for most species, growth and reproduction show seasonal trends.

For species in northern temperate regions, reproduction is limited to the summer and the length of the breeding season shows little difference from year to year. Growth is very rapid in spring as they reach sexual maturity, and the weaned youngsters also grow quickly during the summer, but growth is halted in winter. Reproduction is more variable in tropical species and may occur almost continuously in some if conditions are suitable, although for most species it is timed for seasonal rains and flushes in vegetation and invertebrate prey. As we saw in Chapter 2, shrews have two reproductive strategies with respect to litter sizes and the number of litters produced. Soricine shrews can be considered as r-strategists. These are organisms selected for their high reproductive rates, equipping them well for rapid colonisation and exploitation of habitats when conditions are favourable. These shrews produce a few (two to three) large litters in quick succession during a restricted breeding season, taking advantage of the limited summer period. This produces a sudden increase in the population. Crocidurine shrews are more conservative and less explosive, behaving more like K-strategists in which selection is strong for the production of relatively few, well-developed offspring which have a good chance of survival in competition with others. Crocidurine shrews have smaller litter sizes, the young are slightly larger at birth, and they may breed continuously over an extended period, not synchronously, as in the case of the soricines. The population can then increase gradually and not be subject to such great fluctuations in numbers.

One great advantage of being small and having a high basal metabolic rate is that it permits a faster rate of growth or biosynthesis. Small mammals have much shorter generation times than large mammals. Growth is faster so they have shorter gestation periods, they reach weaning age sooner, and they may even have increased fecundity. For *Sorex araneus* and *S. minutus*, for example, the gestation period is approximately three weeks, and at the time of weaning (about three weeks after birth) the offspring have already achieved a size and weight very close to that of the autumn and overwintering sub-adults. Shrews with the highest rates of production of offspring have the highest basal metabolic rates.

SUMMARY

In summary, two major energetic strategies have evolved in shrews which have permitted them to survive a wide range of environmental conditions. One is an energetically expensive strategy which is typical of the shrews found in temperate and cold environments in northern latitudes. It is exemplified by the subfamily Soricinae, and involves having a very high metabolic rate. Species of *Sorex* have the highest metabolic rates. *Neomys* and *Blarina* show the same trend, but to a lesser extent. Despite the energetic drain of a high metabolic rate, the advantage is that it leads to a greater rate of thermogenesis (heat production) which permits homeostasis to be maintained, even at low ambient temperatures. It also leads to a higher rate of biosynthesis and growth and may permit larger litters to be produced.

The alternative strategy is less expensive energetically and involves a lower metabolic rate. It is best suited to those shrews inhabiting warmer, more southerly latitudes. It is typical of the subfamily Crocidurinae, but *Notiosorex* and, probably, *Cryptotis* show a similar strategy. Combined with an ability to enter torpor it allows homeostasis to be maintained under conditions of heat stress where thermogenesis must be controlled at high ambient temperatures. While rates of growth and litter size may be slightly reduced in these shrews, this may be compensated for by increased longevity.

Accompanying these major strategies are a number of other physiological and behavioural features which are employed variously to assist shrews to overcome problems of energy expenditure. They include increasing thermal insulation with a thick pelage; modification of activity patterns and/or entering torpor; building well-insulated nests and even sharing nests; undergoing a decrease in body weight and reducing absolute food requirements; and, in certain cases, food hoarding.

The main features distinguishing energetic design of the two major groups of shrews have been summarised by Genoud (1988) and are outlined in Table 6.5. The differences between them can be attributed to the evolution of these two groups of shrews in geographical isolation under different climatic conditions.

Table 6.5   The two contrasting energetic strategies exhibited by soricine and crocidurine shrews (modified from Genoud, 1988)

Soricine-type strategy	Crocidurine-type strategy
(for a cold, seasonal climate)	*(for a warm climate)*
Basal metabolic rate high	Basal metabolic rate moderate
Body temperature high, and precisely regulated	Body temperature moderate and more variable
Energy expenditure high	Energy expenditure moderate
Winter reduction in body size	No winter size reduction
No torpor	Torpor
Activity rate high	Activity rate moderate
Large home range	Smaller home range
Solitary	More social
No nest sharing	Nest sharing
Large litter size	Small to moderate litter size

7 Community structure and habitat preferences

SPECIES DIVERSITY AND ABUNDANCE OF SHREWS

Shrews are important constituents of small mammal communities in forests, grasslands and scrublands throughout much of the world. This is reflected in their widespread occurrence, in the high population numbers of some species, and in the diversity of species which exist in a single habitat. An indication of the diversity of shrews occurring in different geographical regions is shown in Table 1.2. Over much of Europe, *Sorex araneus*, *S. minutus* and *Neomys fodiens* frequently coexist in forests and grasslands, and these three species are in places accompanied by *S. caecutiens* and *N. anomalus*. Communities of five species have been studied in France, including *S. coronatus*, *S. minutus*, *N. fodiens*, and the white-toothed shrews *Crocidura russula* and *C. leucodon*. Communities of up to seven species of *Sorex* have been recorded in the Altai-Sayan Mountains of central Asia (Yudin *et al.*, 1979).

The diversity of species in North America is also high, with as many as six or more occurring in a single habitat. For example, in marshland habitats in Manitoba, *Sorex cinereus*, *S. arcticus* and *Blarina brevicauda* commonly occur and these may be accompanied by *S. hoyi*, *S. palustris* and *S. monticolus* (Buckner, 1966a, Wrigley *et al.*, 1979). In tropical regions, communities of shrews are even more complex: 25 different species belonging to five different genera have been reported from part of the Kivu region of Zaire, with 8–10 species occurring together in primary forest and swamp (Dieterlen and Heim de Balsac, 1979).

Despite this richness of species, or perhaps because of it, the abundance of individual species and the contribution of shrews to the total small mammal fauna of an area can be low. For example, in 13 habitats where shrews occurred in central Zaire, they comprised only 4 per cent of the total numbers of captures of small mammals (Dieterlen and Heim de Balsac, 1979). In a mixture of habitats ranging from grassland and scrub to forest in Uganda, only 5.4 per cent of all small mammals caught were shrews, although they contributed 33 per cent of the species richness. However, the abundance of shrews in relation to other mammals may be highly variable, depending upon habitat, season, and even year. In another part of Uganda, the Queen Elizabeth Park, a very different picture of shrew abundance was revealed. Using a mixture of traps in a variety of habitats such as grassland, thicket and forest, Delany (1964) found that shrews (all of the genus *Crocidura*) comprised 22 per cent of the captures of small mammals and 24 per cent of the species richness. In grassland and scrub in southern England, shrews ranged from 28 per cent to 86 per cent of all small mammal captures, with an average of 58 per cent in summer and 49 per cent in winter, over a two-year period of live-trapping. Shrews do not appear to undergo such marked variations in population numbers from year to year as many rodents do, and so in poor rodent years they may dominate the small mammal fauna.

Some of the reported variations in the abundance of shrews in different areas may well be a reflection of research effort and the use of unsuitable or inefficient methods of catching shrews. These small mammals have been the subject of considerable attention for ecologists in northern latitudes, but not so in tropical and sub-tropical regions. It is difficult to gain a true picture of the importance of shrews in small mammal communities since they are notoriously difficult to sample accurately. For example, when snap traps were used, Kirkland (in press) found that 33 per cent of the total catch comprised shrews in south-central Pennsylvania, and 90 per cent of these were short-tailed shrews (*Blarina brevicauda*), a relatively large species. When pitfall traps were used, 58 per cent of all captures were shrews, and *Blarina brevicauda* comprised only 31 per cent of shrews caught. Pitfalls sampled the smaller species more successfully, including the tiny *Sorex hoyi* which had never been recorded during snap-back trapping over many years.

While white-toothed, crocidurine shrews are diverse in terms of species and form the major component of the shrew fauna in tropical regions, their abundance may be low compared with that of red-toothed, soricine shrews which are dominant in temperate regions. Similarly, shrews comprise a smaller proportion of small mammals at high altitudes in tropical and sub-tropical regions, in areas such as Nepal (Abe, 1982) compared with temperate regions.

Table 7.1 Shrews as components of the small mammal fauna in different areas of eastern USA (after Kirkland, in press)

Area	Number of localities	Shrews as % of all small mammal species	Shrews as % of all small mammals captured
Nova Scotia & New Brunswick	8	36.2	29.8
Maine	6	51.2	33.7
New York	21	35.7	30.5
Pennsylvania	14	44.1	18.9
West Virginia	6	40.0	30.5
South Carolina	2	42.9	37.6

Shrews seem to be of greater importance to individual communities of small mammals in northern latitudes, in terms of their abundance. For example, in a study of 57 localities from Nova Scotia southwards to South Carolina, Kirkland (in press) found that shrews comprised an average of 31 per cent of all individuals and 40 per cent of all species of small mammals captured (see Table 7.1). Shrews are major constituents of the forest communities in this region.

Despite the diversity of shrew species in many local areas, they frequently comprise only a small proportion of the small mammal community in most geographical regions, and are usually outnumbered by the many species of rodents such as mice and voles. For example, of the 237 species of small mammal (those with a mean adult body mass of less than 1,000g, including most squirrels and the smaller lagomorphs) in North America, 31 species are shrews. They constitute, therefore, only 13 per cent of the small mammal fauna (Kirkland, in press). Using the same criterion, in Europe the twelve species of shrew constitute 12 per cent of the small mammal fauna, and in the British Isles, the five species of shrew comprise 12.5 per cent. In southern Africa south of the Cunene-Zambezi Rivers (which includes Botswana, Zimbabwe, Namibia, South Africa and part of Mozambique), shrews comprise only 8 per cent of the total small mammal fauna.

Figure 7.1 Regional differences in the importance of shrews in small mammal communities in North America. Shrews are represented as a percentage of the total small mammal species (after Kirkland, in press)

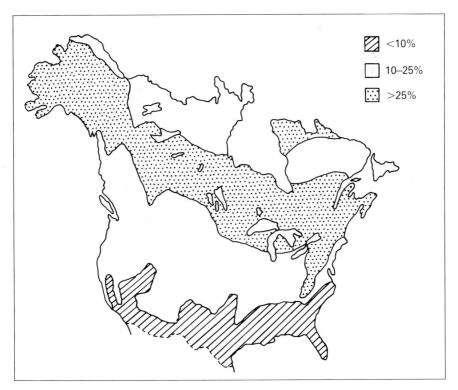

However, there are regional differences in the contribution of shrew species to the small mammal fauna. For instance, in North America they generally constitute less than 10 per cent of the species richness of the small mammal fauna in the desert-scrublands of the south-western USA and the forests of the south-east (see Figure 7.1). They comprise 10–25 per cent in much of the Great Plains and the mountainous areas of the western USA and Canada. But in the north-central and north-eastern part of North America they constitute 25 per cent or more of the species richness (Kirkland, in press). This area embraces the extensive boreal coniferous forests together with the transitional zones between them and the deciduous forests of the east and the grasslands of the mid-west. At very high latitudes in North America, where the mammalian fauna is relatively poor in species, shrews also constitute a major proportion.

These regional differences in species diversity are often associated with the distribution of particular types of shrew. For example, the prolific genus *Sorex*, with some 52 species, is typically an inhabitant of northern latitudes, both in the Palaearctic and Nearctic. *Blarina* (with three species) and *Cryptotis* (with 13 species), have a predominantly southern distribution in the Nearctic. Similarly with *Crocidura*, which is a genus with a multitude of species. It is mostly southern in distribution and it is relatively poorly represented in the Palaearctic. Trends in distribution are also found when traversing different altitudes: the Soricinae occur at higher altitudes, the Crocidurinae only at lower altitudes (Abe, 1982). This is consistent with crocidurine shrews being typically inhabitants of warmer climes.

DOMINANCE AND RARENESS IN SHREW COMMUNITIES

In all shrew communities comprising more than one species, there are marked differences in the relative abundance of the species present. It is usual for one or two species to be dominant in terms of numerical abundance, and others to occur in varying numbers but typically at much lower densities than the dominant species (see Table 7.2). Some shrews in large multi-species communities appear to be very rare. In most of Britain and Europe, for example, it is usual for the common shrew (*Sorex araneus*) to be the dominant species. Pygmy shrews (*S. minutus*) are not often as abundant. In Britain, they represent only 10–18 per cent of all shrew captures in most habitats. Even in Ireland, where they are the only species present and a full range of habitats is available to them, population densities are lower than those of *Sorex araneus* in mainland Britain. Similarly, a community of five species in Uganda had two abundant ones (*Crocidura turba* comprising 53 per cent and *C. flavescens* 22 per cent of captures) while the others were much less numerous, even rare, each contributing as little as 3.7 per cent of captures.

In a study of 47 communities of shrews in forested habitats in eastern North America where more than one and up to six species occurred, Kirkland (in press) found that the most abundant species comprised an average of 67 per cent of shrews captured, while the least abundant species averaged 10 per cent. But although all these communities tended

Table 7.2   Commonness and rareness in shrew communities

Area of study	No. of species in community	Ranked abundance of each species (% of captures)					Source
		1	2	3	4	5	
Northern France	5	69.2	20.5	3.9	3.6	2.7	Yalden et al., 1973
Manitoba, Canada	5	75.3	20.7	4.1	rare	rare	Buckner, 1966a
Uganda	5	53.1	22.2	17.3	3.7	3.7	Delany, 1964
Poland	3	65.7	30.9	3.4	—	—	Aulak, 1967
Southern England	3	52.0	31.0	17.0	—	—	Churchfield, 1984a

to be dominated by a single species, it was not always the same one. In 57 per cent of the communities *Blarina brevicauda* was the most abundant species, in 23 per cent *Sorex cinereus*, and in 19 per cent *S. fumeus*.

Interestingly enough, all three shrews were also the least abundant species in some communities, and so patterns of numerical dominance are not consistent. The reasons for this are not clear, but again there does seem to be a geographic basis to their distribution. *Blarina brevicauda* is typically southern in distribution and so might be expected to be most abundant in the core of its range and less numerous at the periphery, and indeed it was dominant in the southerly part of the range sampled by Kirkland. By contrast, *Sorex cinereus* is typically boreal in distribution and it predominated in the more northern communities.

FACTORS AFFECTING THE DISTRIBUTION OF SHREWS

The reasons for the variation in the regional abundance and diversity of shrews may lie in the characteristics of the habitats in diffferent areas, and the resources and niches available. The abundance and diversity of shrews in the cool, damp boreal forests of North America may be due to the abundant, diverse and dependable food resources found there in the form of soil invertebrates. This may permit the coexistence of many different types of shrew and reduce competition for resources. In other areas where there may be periodic droughts and seasonal shortages of food, such as in the drier oak forests of eastern North America, there may be fewer niches available, and the number of shrew species may be limited by competition for food.

Shrews do tend to be most abundant and diverse in regions character-ised by cool, moist, temperate forests. They are less diverse in drier forests and prairie or steppe ecosystems, and are least diverse and abundant in deserts. Moisture, then, must be a principal factor in determining the regional and local diversity of shrews. A comparison of the abundance and diversity of shrews in wet (hydric), moist (mesic) and dry (xeric) habitats in Manitoba, Canada supports this (Wrigley et al., 1979). Shrews were more numerous and had a higher diversity of species in the wetter sites. They may be excluded from the drier habitats and show a preference for moister areas. This is consistent with the fact that shrews have relatively high water requirements: they have a high evaporative respiratory loss compared with small rodents and seem unable to regulate evaporation at low humidities. This may be due to their high metabolic rates and their constant high level of activity. So dessication may be a problem for them and they may need moist habitats with access to free water to overcome this (see Table 7.3).

Table 7.3   Distribution of shrews in wet and dry habitats in Manitoba (after Wrigley et al., 1979)

Habitat type	No. habitats	% of shrews in each habitat type	Average no. species
Hydric	6	49	4.7
Mesic	14	46	3.5
Xeric	10	5	1.8

However, moisture, particularly in the form of rainfall and fluctuating water tables, can have an adverse effect on those species which exhibit a preference for wetter habitats. In *Sorex cinereus*, which occurs in damp sites, there is evidence that rainfall adversely affects nesting and increases the mortality of juveniles.

Probably just as important to shrews in determining their abundance and diversity is the effect that water has on the biomass, abundance and diversity of their invertebrate prey. Many invertebrates are highly influenced by environmental moisture, which may not simply affect their presence or absence in an area but also the form of their life cycles. Prolonged periods of drought may be overcome by aestivation or quiescence deep in the soil when they may be inaccessible to shrews. Certainly invertebrate availability is greatest in damper sites, and a positive relationship between the abundance of soil invertebrates and shrews has been found. But is this reflected in the diversity of shrews? The productivity hypothesis of species diversity (MacArthur, 1965, 1972; Pianka, 1971) proposes that the more productive habitats will support a greater number of species because of the opportunities provided for subdivision of resources and specialisation. So, where prey are diverse and abundant, more niches are available for predators, which permits species packing and a greater species richness. The small size of shrews may also be a contributory factor since they can exploit their habitats in a more course-grained fashion than is possible amongst communities of larger species (MacArthur and Wilson, 1967).

Plant cover does not seem to be an important factor directly affecting the distribution of shrews except perhaps by the provision of protection against predators. However, it has a very important indirect effect due to its influence on general conditions of humidity and on invertebrate numbers. The leaf litter cover provided is also important to invertebrates and for the hunting tactics of many shrews.

In a study of the factors affecting the habitat distribution of *Blarina brevicauda* and *Sorex cinereus* in southern Michigan, Getz (1961) considered a variety of environmental factors including temperature, moisture, vegetation type and availability of cover and food, as well as interspecific competition. He concluded that temperature, type of cover and interspecific competition were not important factors directly affecting habitat selection by these two shrews. But *B. brevicauda* was only found in moist areas and was absent wherever large invertebrates were low in availability. *S. cinereus* was found in all habitats, except upland hardwoods, but there was no correlation between its occurrence and the availability of large invertebrates. Being a diminutive shrew, it may be capable of utilising small invertebrate prey such as spiders and ants more effectively than can the larger *B. brevicauda*, and so may not be so constrained in its use of habitats. He surmised that food availability was the most important factor in the distribution of shrews, but this in turn is affected by moisture and vegetation cover.

Interspecific competition is thought by many to be an important factor affecting the distribution of a species. The failure of similar species to coexist and have overlapping ranges may be caused by differences in habitat requirements, but it may be the result of mutual exclusion through the effects of competition between these species.

An interesting situation arises in Europe with two very closely related shrew species, the common shrew (*Sorex araneus*) and Millet's shrew (*S. coronatus*), which has only recently come to light with the work of zoologists in Switzerland, particularly Meylan and Hausser (1978) and Neet and Hausser (1990). These two species are almost identical— so much so that they cannot be reliably differentiated in the field. But they do differ genetically, and do not interbreed. While *S. araneus* is widely distributed over much of central and northern Europe, including Germany, Austria, Poland, Hungary and Czechoslovakia (and most of the British Isles), *S. coronatus* is found primarily in France and northern Spain. There is a broad contact zone in western Germany and Switzerland where both species occur, and their ranges may overlap. In Switzerland, where the distribution of these species has been studied in most detail, there are small areas where they have overlapping ranges, plus areas where only one or the other occurs. This provides an excellent opportunity to examine the local distribution of the two species, their habitat selection, and the role of competition in their distribution.

Such a study was carried out by Neet and Hausser (1990) in the contact zone between the two species in Switzerland. They looked firstly at habitat use by the two species to see if the segregation between them might be due to different habitat requirements or preferences. They then went on to perform some removal experiments to investigate the effect on one species of removing the other.

An examination of the spatial distribution of capture locations in the areas where both species occurred revealed a localised patchwork of individual territories, with no distinct barriers between the species. However, exhaustive examination of the characteristics of the habitat in which each species was captured (including the nature of the vegetation, the soil type, the depth of leaf litter, the organic content of the soil and humidity) did show that where the two species coexisted there was some habitat segregation. *S. araneus* occupied more humid habitats with a thicker litter layer than *S. coronatus*.

The removal experiments also produced some interesting results. Figure 7.2 shows the effects of removing either *S. araneus* or *S. coronatus*, by successive trapping, on the population size of the other species. Three out of the four experimental plots exhibited an increase in the numbers of the unmanipulated species when the competitor species was removed. Although the numbers involved were small, and the results somewhat equivocal, the results did suggest that the species were affecting each other in some way. Moreover, Neet and Hausser concluded that habitat segregation was a consequence of interspecific interactions between the species, because the pattern of habitat segregation prior to the removal experiment disappeared during the experiment. With the absence of the competitor species, the remaining species broadened its habitat distribution and widened its niche to incorporate habitats previously occupied by the competitor. They concluded that habitat segregation allows *S. araneus* and *S. coronatus* to tolerate local overlap and coexistence in the contact zones of their range. So the shrews seem to behave differently according to the presence or absence of the competitor species, although exactly how this operates remains unknown.

Figure 7.2 Changes in population numbers of Sorex coronatus *or* S. araneus *following the removal of the opposite species. Vertical dots represent removals (after Neet and Hausser, 1990)*

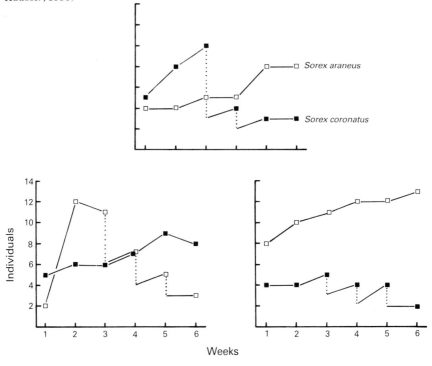

HABITAT PREFERENCES

While many species of shrew may overlap in range, they may have slightly different habitat requirements or tolerances. Hence the local distribution may be affected by habitat selection by different species of shrew which itself may reduce contact and competition between species. In the case of *Sorex araneus* and *S. coronatus*, as we have just seen, habitat selection appears to be very subtle with humidity and organic content being important criteria. In other cases, there may be more obvious differences in habitat preferences. For example, shrews in northern France occur widely in a variety of habitats. Different species exhibit certain preferences, but these are not exclusive (Yalden *et al.*, 1973). The dominant species, *S. coronatus* (previously thought to be *S. araneus*), was widespread amongst different habitats but was mostly found in grassland and marshland. The pygmy shrew, *S. minutus*, showed a clear preference for grassland and stone-wall habitats, and this was mirrored by the white-toothed shrews, *Crocidura leucodon*. The water shrew (*Neomys fodiens*) inhabited mostly marshy areas and pond edges, and also grassland; and the white-toothed shrew *C. russula* was found mostly associated with stone walls and around inhabited buildings in a commensal situation (see Table 7.4).

Table 7.4 Habitat preferences of shrews: the percentage of each species captured in different habitats in northern France during three summer sampling periods (after Yalden et al., 1973)

	Sorex coronatus	Sorex minutus	Neomys fodiens	Crocidura russula	Crocidura leucodon
Woodland	7.9	0	0	10.3	15.4
Wood edge	10.5	0	0	5.9	15.4
Grassland	43.7	33.4	41.6	10.3	30.8
Marshland	21.8	11.1	41.6	4.4	0
Stone walls	7.4	44.4	0	17.6	30.8
Around buildings	8.7	11.1	16.8	51.5	7.6

Comparing the habitat preferences of a community of five species of shrew in a mixture of vegetation types in Colorado, USA, Spencer and Pettus (1966) also found evidence of quite distinct difference between species. The small shrews *Sorex hoyi* and *S. nanus* occurred relatively more frequently in coniferous forest than the other species. This might reflect their superior ability to exploit small arthropod prey. The larger *S. cinereus* and *S. vagrans* were found in damper sites, mostly in marshland; and the water shrew (*S. palustris*) was found only in marshy habitats adjacent to water.

Little information is available about the habitat preferences of shrews in tropical and sub-tropical areas, but some species appear to be dominant and ubiquitous while others are more restricted in occurrence, as in temperate regions. For example, in 18 different habitats characterised by vegetation cover and altitude in the Natal Drakensberg of South Africa, two species of shrew were found. The mouse (or forest) shrew, *Myosorex varius*, was numerically dominant and had the widest distribution. It occurred in all vegetation types from grassland and scrub to forest, and at all altitudes. The white-toothed shrew *Crocidura flavescens* was more restricted, occurring only in grassland areas and not in woodland or forest (Rowe-Rowe and Meester, 1982).

Moisture may be an important factor affecting the habitat distribution of shrews in certain parts of Africa where there are marked wet and dry seasons. While several species are tolerant of dry conditions in grassland and bush and so are widespread in distribution, others are restricted to moister habitats such as marshes and are not as ubiquitous. Table 7.5 shows the distribution of shrews in the Orange Free State, South Africa (Lynch, 1983). Two species of *Suncus* show a particular kind of habitat association: they have mostly or exclusively been found in disused termite mounds where they make their nests. In dry grassland, where cover may be limited and burrowing difficult, termite mounds with their multitude of galleries both above and below ground provide a good refuge for these small shrews. Conditions there may also be cooler and moister than the surrounding grassland, and attract invertebrate prey.

Table 7.5 Habitat preferences of shrews in the Orange Free State, South Africa (after Lynch, 1983)

Species	Common name	Habitat preference			Distribution
		Wet/dry	Situation		
Crocidura mariquensis	Swamp musk shrew	Wet	Marshes		Localised
Myosorex varius	Mouse shrew; forest shrew	Moist	Marsh; dense grass & shrubs		Localised
Crocidura hirta	Lesser red musk shrew	Moist	Marsh, forest, grass-scrub		Highly localised
Crocidura bicolor	Tiny musk shrew	Moist/arid	Grassland, vleis, bush. Diverse		Widespread
Crocidura cyanea	Reddish-brown musk shrew	Dry	Rocky koppies, grass, shrubs		Widespread
Suncus infinitessimus	Least dwarf shrew	Dry	Termitaria; grass & shrubs		Highly localised
Suncus varilla	Lesser dwarf shrew	Dry	Termitaria; open grassland		Widespread

In addition to moisture and the availability of cover, the amount of disturbance may be an important factor in habitat selection. In a study of the habitat use of small mammals in a variety of habitats, including grassland, thicket and forest, in the Queen Elizabeth Park, Uganda, Delany (1964) found that 66.7 per cent of all shrews captured were in undisturbed grassland not heavily populated by large herbivores, compared with only 3.7 per cent in heavily grazed and periodically burned grassland and 11 per cent in grazed grass and thicket. The remainder occurred in relatively undisturbed scrub and forest.

Some shrews habitually show a preference for particular habitats. The water shrews *Nectogale elegans* and *Chimmarogale* are reported only to occur beside mountains streams in Asia, and are the best-adapted of all shrews for a semi-aquatic lifestyle. The European and American water shrews (*Neomys fodiens* and *Sorex palustris*) are not quite so restricted, but still show a preference for streamsides or wet, marshy habitats. While most shrews prefer moist habitats, there are a few species which habitually inhabit very dry steppe or semi-desert, for example the piebald shrew *Diplomesodon pulchellum* in central Asia, and the desert shrew *Notiosorex crawfordi* in the south-western USA. *Notiosorex* actually inhabits a variety of dry habitats especially semi-desert scrub comprising mesquite, agave and scrub oak where it nests on the surface of the ground beneath dead or dying agaves or piles of brush or in woodrat houses. It is a relatively rare species of shrew with low population densities, probably as a result of poor prey availability. Little is known about this small shrew and how it survives in such a rigorous environment.

Habitat preference of a rather unusual kind is exhibited by the white-toothed shrew, *Crocidura russula*, in Europe. In the northern part of its range, and in the mountainous areas of central and western Europe, this shrew is restricted to the vicinity of human dwellings. In a study of these shrews in a rural mountain habitat in Switzerland, their local distribution and behaviour was found to vary considerably through the year (Genoud and Hausser, 1979). In summer they occurred in areas of dense herbaceous and shrub cover and were mainly active at ground level amongst the leaf litter. But, as autumn came, conditions became colder and the vegetation began to die off, the shrews became more active around and inside compost heaps and buildings where the microclimate was milder and invertebrate prey more abundant. None of the other shrew species in the area showed this trend and, indeed, were rarely encountered in the vicinity of human dwellings. *C. russula*, which originated in warmer, more southerly regions, is not well suited to a cold climate but, being an opportunist, it takes advantage of more equable conditions wherever it can find them.

The inability of both *C. russula* and *C. suaveolens* to thrive in cold climates may be another reason why they are not found in mainland Britain. It seems likely that they could have been introduced to the mainland by the same accidental process which apparently brought them to the Channel Islands and the Scillies, but they were only able to survive in the warmer conditions of these small islands in the extreme south of Britain.

SHARING HABITATS

Although many species of shrew may have similar habitat requirements, in the form of vegetation cover and invertebrate food resources, and seem at first sight to be extremely similar in habits, there is growing evidence that different species may use their habitats in different ways. This is particularly so where several species coexist in the same habitat, but it often seems to occur in quite subtle ways. Vertical segregation forms the basis of differential habitat use in a variety of shrew communities.

For example, in many terrestrial habitats in Britain and Europe, both *Sorex araneus* and *S. minutus* coexist. However, field studies show that *S. minutus* is nearly always outnumbered by *S. araneus*. Moreover, *S. minutus* also has a larger territory/home range than *S. araneus*, despite being a smaller shrew. Does competition between the species influence their habits in any way, particularly those of the smaller species?

Field studies by Michielsen (1966) in the Netherlands suggested that competition between *S. araneus* and *S. minutus* has resulted in slight differences in niche occupancy by these species. Her programme of live-trapping resulted in a theory of spatial separation between the two species, at least in autumn and winter, because there were times when *S. minutus* was easily caught on the ground surface but *S. araneus* was not. Michielsen concluded that *S. minutus* was most active on the surface of the ground (epigeal) while *S. araneus* was most active underground (hypogeal). This partial vertical segregation thus reduced competition and interference between the two coexisting species.

Further studies have confirmed that there does indeed seem to be partial segregation between these species. Individually marked common shrews frequently 'disappear' from the population for weeks or months on end in late autumn and winter when they become untrappable. They re-emerge some time later to be recaptured at the same sites once again (see Figure 7.3).

Additional evidence of their partial segregation has come from studies of their feeding ecology. *S. minutus* feeds predominantly on prey found on the ground surface while *S. araneus* includes a large proportion of soil invertebrates (such as earthworms and certain dipteran larvae) in the diet and a relatively smaller proportion of prey from the ground surface (see Chapter 5). Michielsen suggested that *S. minutus* has to maintain larger territories because its prey are less abundant, and perhaps less

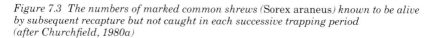

*Figure 7.3 The numbers of marked common shrews (*Sorex araneus*) known to be alive by subsequent recapture but not caught in each successive trapping period (after Churchfield, 1980a)*

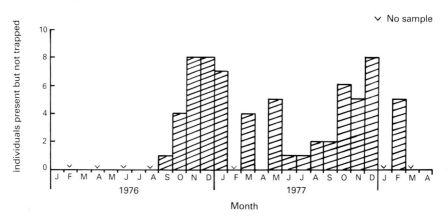

142

clumped in distribution, than soil-dwelling invertebrates fed on by *S. araneus*, but this has yet to be verified.

Studies of other shrew communities have revealed similar trends in differential habitat use. For example, in Japan, two species of shrew are common and ubiquitous, and they often co-exist in the same habitat. *Sorex unguiculatus* is a relatively large species which possesses long claws on the fore feet to assist in digging. It burrows well and feeds primarily on earthworms. *S. caecutiens saevus*, which often shares the same habitat, is smaller with only short claws on the fore feet. It is not such a good digger, but instead spends most of its time on the ground surface where it feeds on spiders. Vertical segregation operates, therefore, with one species being primarily subterranean while the other lives mostly on the ground surface. Moreover, although both species are active by day and by night, *S. caecutiens* is more nocturnal than *S. unguiculatus* (Yoshino and Abe, 1984).

In a study of small insectivores in Tsuga forests in Washington State, north-western USA, Terry (1981) showed that different species used the same habitat, but in slightly different ways. The shrew mole *Neurotrichus gibbsi* occupied forested areas where it burrowed deeply and was primarily subterranean. The shrew *Sorex trowbridgii* was also capable of burrowing, but not so deeply. It occurred in areas of good ground cover where it burrowed in the surface, organic soil layer. *S. monticolus* had a fairly uniform distribution but was not a burrower. Instead, it foraged amongst the leaf litter on the forest floor. *S. vagrans* was also not a burrower, and had a more patchy distribution. It preferred more open, damp areas where the water table was high and there were few *S. trowbridgii*.

Small differences in adaptation for a particular mode of life lead to quite subtle ways of partitioning habitats and their resources between species. For example, *Sorex obscurus* is able to maintain territories against *S. vagrans* in acidic, western hemlock habitats while *S. vagrans* has a competitive advantage in red cedar habitats. The stability of this relationship between the two species over several years of study suggests that there may be a basic difference between them (Hawes, 1977). *S. obscurus* has been found to have rather more robust teeth which wear down more slowly than those of *S. vagrans*, and this could be an adaptation for feeding predominantly on small, highly chitinous arthropods which are particularly abundant in acid soils. In the richer, less acidic habitat where *S. vagrans* occurred, larger insects and their larvae were relatively more abundant and they were what it fed on. So *S. obscurus* may be better adapted to the more acid sites where its more durable teeth, possibly coupled with a different foraging strategy, make it competitively superior to *S. vagrans*. In the richer sites preferred by *S. vagrans*, the greater biomass of food may permit the higher reproductive effort observed in this shrew, allowing it to be competitively superior in this habitat.

Differences in the ability to exploit habitats and the resources they provide may also form a basis for the colonisation of certain areas by some species but not by others. Some species, such as *S. minutus*, are euryoecious, having a wide ecological amplitude and occupying diverse habitats. Others are stenoecious, being more restricted in distribution, such as *S. alpinus* and *S. dispar* which occupy mainly rocky habitats and mountain stream sides. The pygmy shrew (*Sorex minutus*) is relatively more numerous in both moister and drier habitats than the common shrew (*S. araneus*), and has a better resistance to drought and flooding. The common shrew seems to be more prone to flooding, probably because it spends a considerable time in underground tunnels. Such differences may have a long-term and interesting biogeographical outcome. For example, there has been considerable speculation over why the pygmy shrew occurs in Ireland but the common shrew does not. Since the pygmy shrew is greatly outnumbered by the common shrew wherever they are sympatric and syntopic in their range in Britain and Europe, this situation is even more difficult to explain.

Some light has been cast on the subject by Yalden (1981) who observed that in areas of moorland with a deep covering of peat, the pygmy shrew can outnumber the common shrew. This was related to a major dietary difference between the two species: earthworms form a major part of the diet of the common shrew but not the pygmy shrew. Earthworms are not found in peat where the acidity and the tendency to waterlogging are not conducive to them. Hence, moorland areas are more suited to the predominantly surface-dwelling, arthropod-feeding pygmy shrew than the subterranean, earthworm-eating common shrew. There is evidence of a low-lying, probably partially flooded, land bridge between Scotland and Ireland around 8,600 years ago, which persisted for about 1,000 years. This may have allowed the pygmy shrew to survive there and eventually to migrate to Ireland.

COMPETITION AND CHARACTER DISPLACEMENT

The best way of investigating competition and its effects on similar, coexisting species is to study the species together (in sympatry) and again on their own (in allopatry), and look for differences between them.

Where two similar species have overlapping distributions in some areas but not in others (such as *Sorex araneus* and *S. minutus*) they may show greater differences in the area of overlap than in the area of allopatry. These differences come about through evolutionary change in response to interspecific interactions such as competition, and the phenomenon is known as character displacement. In allopatric coexisting species, character displacement is reflected in differences in their habits (including diets and foraging strategies) and their physical attributes, such as body size or jaw length. Looking for signs of character displacement provides an indirect method of examining competition between species. In studies where this has been done, some interesting but rather equivocal results have been obtained.

A good place to investigate the habits of pygmy shrews in isolation is Ireland where, as we have seen, they occur in the absence of their potential competitor, the common shrew. Ellenbroek (1980) made a study of the population density, territory size and surface activity (the 'epigeal-ness') of pygmy shrews in different sites in Ireland, and compared them with those of pygmy shrews in similar habitats in the Netherlands where *S. araneus* also occurred. He did this by employing live-trapping techniques coupled with a novel method of recording the shrews' activity and movement patterns on the surface. He placed sheets of smoked paper in the bottoms of shallow boxes or trays. Lids were placed on the trays and small holes made in the sides to permit the entry and exit of the shrews (see Figure 7.4). The trays were placed on the ground, distributed around the home ranges of shews. As they entered and explored the boxes, they left tell-tale signs of their presence in the form of foot-prints etched into the smoke-blackened surface of the paper.

Figure 7.4 Smoked-paper box used to study the activity of pygmy shrews (Sorex minutus) *on the ground surface (after Ellenbroek, 1980)*

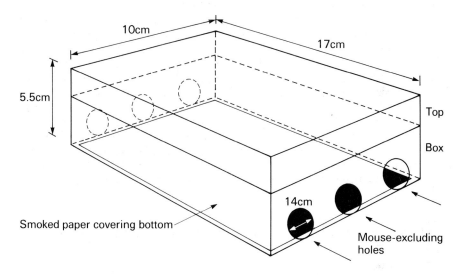

The prediction was that if competition did occur between *S. minutus* and *S. araneus* then, in the absence of the larger, more numerous species, *S. minutus* should be able to modify its habits to make greater use of its habitat. Hence, in allopatry, it should be able to have smaller territories, larger population densities and occupy both epigeal and hypogeal niches. However, the results were not as Ellenbroek predicted, and he found no significant differences in habits between his Irish and Dutch populations of pygmy shrews.

So perhaps competition is not an important factor affecting the habits and niche occupancy of these shrews. An alternative explanation may be that *S. minutus* has become permanently adapted by evolutionary change to occupy a particular niche (epigeal) so that, even in the absence of a competitor, it does not change its habits. However, another and perhaps more likely explanation is that the study sites chosen may not have been strictly comparable. Although they may have looked similar, to a shrew they may have had different characteristics in terms of environmental conditions or food resources, which contributed to the findings.

Differences in distribution between the common and the pygmy shrews were also used by Malmquist (1985) as a means of investigating competitive displacement in these species. He studied the two species in sympatry on the mainland of Sweden and Scotland, and in the Orkney Islands, and the pygmy shrew in allopatry in Ireland, the Outer Hebrides and Gotland. Allopatric common shrews were studied on the small island of Rodloga in the Baltic Sea. The prediction was that pygmy shrews in areas of sympatry would be smaller than those in areas of allopatry. Similarly, if competition affected *S. araneus*, it should be smaller in areas of sympatry. Measurements of the skulls and jaws of the shrews were used as an indication of the presence or absence of character displacement (and hence of competition) in these populations. Differences in jaw size, for example, would suggest an adaptation to prey of different sizes and so a reduction in competitive interactions.

Malmquist took eight different measurements of each skull, and he found, as he had predicted, that *S. minutus* did have significantly larger jaws where it occurred alone in Ireland, Gotland and the Outer Hebrides than where it overlapped in range with *S. araneus* on the Scottish and Swedish mainlands and the Orkney Islands (see Figure 7.5). There was no difference in skull size between the sympatric populations of pygmy shrews from the Orkneys and the mainlands. So the larger size of these shrews in Gotland, Ireland and the Outer Hebrides was not the result of a general insular effect on the populations. Nor did he find any evidence of latitudinal trends in the skull measurements of the pygmy shrews.

So, in the absence of competition from *S. araneus*, the pygmy shrew was able to broaden its niche and eat a wider range of prey sizes, and this

*Figure 7.5 Relative jaw sizes of populations of pygmy shrews (*Sorex minutus*) in sympatry with* S. araneus *and in allopatry. Four different skull measures (A–D) are shown. Vertical lines denote 95 per cent confidence limits (after Malmquist, 1985)*

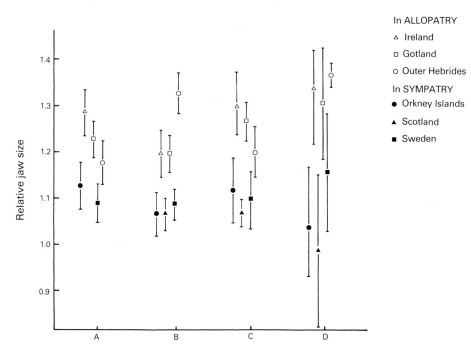

was reflected in the larger jaw size. Where the two species occurred together, *S. minutus* was outdone by the larger *S. araneus*, its niche was squeezed, and so it concentrated on smaller prey, leading to a reduction in jaw size. There were no such differences, on the other hand, between sympatric and allopatric populations of *S. araneus*. This was interpreted as an asymmetry in the competitive effects of these two shrews on each other, with the larger, more numerous and dominant common shrew being competitively superior to the pygmy shrew.

PREDATION AND SHREW COMMUNITIES

Predation is an important component of the dynamics of communities. We have considered the effects of shrews as predators in some detail, but what about the impact of predators on populations of shrews? While many avian and mammalian predators eat shrews occasionally, none include them as a major component of their diet. As we have seen in Chapter 2, owls are the major cause of shrew mortality by predation, but only 5–13 per cent of their diet comprises common shrews. All avian and mammalian predators of small mammals specialise in rodents such as voles and mice. Gosczynski (1977) found that over 60 per cent of the diet of buzzards, barn owls, long-eared owls, weasels and foxes hunting in woodland and farmland in Poland comprised voles (*Microtus*) alone. Even when numbers of voles were low they still featured prominently in the diets of these predators, especially the birds, which were able to hunt out these prey even at low population densities.

This may reflect the greater palatability of rodent prey. But it is more likely to indicate the greater abundance of small rodents compared with shrews, not only in terms of species richness, but also in population density. Hence studies concentrate on the impact of predators on rodent populations, and vice versa. There have been few attempts to quantify the effects of predation on shrew communities. Nevertheless, predation must have an impact on shrew communities. For example, Pearson and Pearson (1947) reported that owls caught younger short-tailed shrews (*Blarina brevicauda*) than they themselves did in their traps, suggesting that there was selection of young animals. As we have seen in Chapter 2, young, inexperienced shrews leaving the parental nest seem to suffer the greatest mortality. It would be interesting to know more about the predators which eat the large, highly odoriferous shrews found in Africa and Asia, such as *Suncus murinus* and the large *Crocidura* species.

8 Shrews and man

MYTHS AND SUPERSTITIONS

Since ancient times shrews have been the source of many different myths and superstitions, and they have gained a poor reputation. The Romans believed that shrews were evil, poisonous creatures, and their reputation has not improved in succeeding ages. In the seventeenth century, the natural historian Topsell described the shrew as 'a ravening beast, feigning itself gentle and tame, but being touched it biteth deep, and poisoneth deadly. It beareth a cruel mind.' The Latin name of *araneus* (meaning spider) was given to the common shrew because it was thought to be poisonous like a spider when it bites its prey. If horses and cattle were put out to feed in pastures inhabited by shrews, they would surely die. If a shrew walked over a sleeping cow it was believed to cause lameness or even paralysis. Antidotes and charms were developed to protect people and their domestic animals from the evil effects of the shrew. One such recipe was to pack the body of a shrew in lime, leave it until it had dried out and hardened, and then hang it around the neck of a cow or horse. This would keep the animal safe from the bite of a shrew.

Shrews also had a reputation for being bad-tempered and ferocious. This must have led to the shrewish character so often portrayed in literature, and exemplified in Shakespeare's play. Why women should be singled out as being 'shrews' can only be speculated upon. Shrews were known to be pugnacious and to fight each other readily. They were also known to put up a brave and fierce defence when cornered by predators, dogs or even humans.

By contrast, shrews were also seen as feeble, highly strung beasts which were easily killed by shock. They could not cross the path of a human being and live, and the mere sight of a human would cause them to drop down dead. They were thought to die if they got into cart tracks, being too weak and feeble to climb out of the ruts. Early naturalists and writers reported that shrews died quickly when they were captured, probably from the fright of being handled. Many agreed that even a small noise was sufficient to cause death from shock. Observations of captive shrews suggested that they were extremely weak and feeble, and that they simply stumbled or tottered about. These early observations were probably made on shrews which were close to death by starvation, since it is unlikely that these early naturalists were aware of their large food requirements, or of what to feed them on, and the symptoms described closely resemble those of starving shrews.

Today, we are a little better informed. Shrews are generally robust, hardly little creatures which do not die readily from shock or loud noises. It is true that certain species, such as *Neomys fodiens*, do produce a mildly narcotising saliva which could pass as a poisonous bite. It is more likely that their reputation for being poisonous derives from early observations that many carnivores, including domestic dogs and cats, will catch but not eat shrews. But this is due to their unpalatable taste and smell rather than to any poison produced by them.

SHARING HOMES WITH SHREWS

Most people rarely come into contact with shrews, and may only see them when they are brought home by the domestic cat. But there are occasional reports of live shrews being found in garden sheds and houses, even attics, having apparently got there by their own efforts. In Britain, these are mostly pygmy shrews, but common shrews are sometimes reported too. Pygmy shrews are agile, they are good climbers and they may make their way into attics while on foraging expeditions amongst creepers or bushes growing up the sides of houses. Once arrived, they could survive for a while on spiders, beetles, flies, silverfish and other invertebrates lurking there.

Some species of shrew have become accustomed to living in close association with man, around human dwellings, barns, gardens and vegetable plots. The white-toothed shrew, *Crocidura russula*, is sometimes known as the house shrew since it is quite often found around human dwellings in Europe and the Near East. Other species of *Crocidura*, including the large *C. poensis*, are found around houses, gardens and rubbish tips in Africa and Asia where they perform a useful function by eating cockroaches, ants and other undesirable insects. They are also known as musk shrews, along with members of the genus *Suncus*, because they produce a distinctive musky odour, and their presence can often be detected more by this odour than by actual sightings of individuals. The large musk shrew *Suncus murinus*, found principally in eastern Asia, has similar habits. It, too, is smelly. It is also noisy, emitting incessant, shrill, clattering sounds as it goes along. It is known by the Chinese as the money shrew because its chatterings are thought to resemble the jingling of coins. But despite its smell and noise, and the fear that it may cause some damage to stored products, it is not regarded badly by local people, as it does destroy many insect pests and possibly even rodents.

Man seems to have been responsible, albeit unwittingly, for introducing certain species of shrew to new lands where they have successfully colonised and established themselves. While this has not occurred on the same scale as the accidental introduction of other mammals such as rats and mice, there are some quite remarkable instances of shrews being transported considerable distances by boat, and surviving the long voyages. For example, the white-toothed shrew, *Crocidura suaveolens*, does not occur on the British mainland but it is found in the Isles of Scilly. It is thought to have been introduced accidentally by traders from France or northern Spain in the Iron Age (or even earlier), as they travelled to the Cornish coast in search of tin. These shrews are better able to cope with food deprivation than *Sorex* species, and so may be able to survive a few days with little food. Similarly, the irregular distribution of both *C. suaveolens* and *C. russula* in the Channel Islands remains difficult to explain except by human introduction. Both species are widespread in mainland Europe, but *C. russula* occurs only on Alderney, Guernsey and Herm of the Channel Islands, and *C. suaveolens* on Jersey and Sark.

The presence of the pygmy shrew on remote islands such as Lewis, Barra and Orkney around the northern British Isles also suggests accidental introduction by man. This species is found all over the Scottish mainland and most of the larger islands close to the coast, so it could have been transported across by boat amongst firewood, fodder for stock or some similar means. How these small shrews survived such a journey is not known, but the smaller absolute food requirements of *Sorex minutus* compared with the larger *S. araneus* make it a better candidate for such voyages. Indeed, there are no known instances of similar introductions of common shrews.

More recently, probably in historic times, *Suncus murinus*, whose natural range is centred on eastern Asia, has been repeatedly introduced to a variety of locations as far-flung as Guam, New Guinea, Madagascar, East Africa and Egypt.

SHREWS AND DISEASE

While shrews harbour a host of different internal and external parasites, they have not been found to suffer widespread mortality due to any viral or bacterial infection as is the case with, say, rabbits and myxomatosis. Yet many viral and bacterial parasites causing diseases of man and his domestic animals have been isolated from shrews as well as from other small mammals. Since shrews are widespread and common, do they present a risk to man in terms of disease transmission?

Tuberculosis has been reported in shrews, including *Sorex araneus* in Britain, where lesions have been found in the internal organs such as lungs, spleen and liver. However, the incidence of tubercular lesions in shrews seems to be very low and is unlikely to cause a problem to man or his domestic animals.

Several species of *Leptospira* have been isolated from shrews, including *S. araneus*, and it has been postulated that shrews may serve as reservoir hosts in the transmission of the bacterial disease leptospirosis to man and domestic stock. *S. araneus* has also been found to carry the viruses causing louping ill in sheep, and the fungus *Trichophyton persicolor* which causes ringworm. Both *S. araneus* and *Neomys fodiens* may act as reservoirs for *Pneumocystis carinii*, the causative agent of a group of infantile pneumonias affecting man.

In Nigeria, a rabies-related virus (Mokola virus) has been isolated from species of *Crocidura*, but it is not though that shrews present a problem either as a reservoir or transmitter of rabies.

Shrews carry a range of fleas and ticks, many of which are responsible for transmitting diseases to man, including rat fleas such as *Xenopsylla cheopis* which carry the bacillus *Yersinia pestis* causing plague. While plague is primarily a disease of rats, other mammals can act as reservoirs, and the disease is transmissible to man by the bite of fleas. There have been several cases in south-east Asia (including Cambodia and Java) and China where the musk shrew *Suncus murinus* has been found to harbour *Xenopsylla cheopis*, and be carrying the plague bacillus. This shrew has been cited as a carrier and potentially serious reservoir of the disease, particularly where it lives commensally with man around villages when it comes into contact with rat infestations. However, although the bacillus has been isolated from these shrews, none have yet show signs of active infection.

Tick-borne encephalitis, a viral disease transmitted by the bite of ticks, has been recorded in populations of *S. araneus*.

Although shrews may have the potential for the transmission of some serious diseases to man or his domestic animals, the risks they present are very low indeed. Unlike rodents such as mice and rats, shrews rarely occur in very close association with man and are unlikely to come into close enough contact to permit disease transmission. Moreover, shrews never have such large, dense populations as are often found in these rodents. Even in species of *Crocidura* and *Suncus* which may live habitually around human dwellings, outhouses and rubbish tips, their numbers are small. The social organisation and behaviour of shrews is such that they do not live in large colonies, but as individuals with relatively little contact with their fellows. This is not conducive to the spread of disease.

SHREWS AND PEST CONTROL

Far from being an evil influence, shrews have a valuable role in the ecosystem, and are highly beneficial to man. This is a result of their predatory activities and their voracious appetites. Studies by Holling (1959) suggested that small, insectivorous mammals such as shrews may have an important role in regulating prey populations, including pest species like European pine sawfly (*Neodiprion sertifer*) which causes widespread damage to conifer plantations in North America. The larvae of this pest feed on conifer leaves, but then drop to the forest floor to pupate in the soil. Shrews and other small mammals feed on the developing pupae or cocoons.

Holling worked in plantations of Scots and jack pine in south-western Ontario where the three most abundant small mammal predators were the masked shrew (*Sorex cinereus*), the short-tailed shrew (*Blarina brevicauda*) and the deer mouse (*Peromyscus maniculatus*) whose populations he studied by live-trapping. He studied the densities of sawfly cocoons by extracting them from samples of leaf litter and surface soil around the trees. He could tell on examination how many cocoons were intact and uneaten, and how many had been opened by small mammals. Species-specific differences in the opening of cocoons allowed Holling to determine which species of small mammal had eaten the prey.

With estimates of the numbers of predators, the cocoons available and the numbers of prey destroyed, he calculated the daily number of prey consumed per predator at different cocoon densities. He found that the number of cocoons opened by each species of predator increased with increasing cocoon density until a maximum daily consumption was reached. This he called a functional response of the predator to increasing availability of the prey, and it is illustrated in Figure 8.1.

Figure 8.1 Functional response of small mammal predators to increasing numbers of prey (after Holling, 1959)

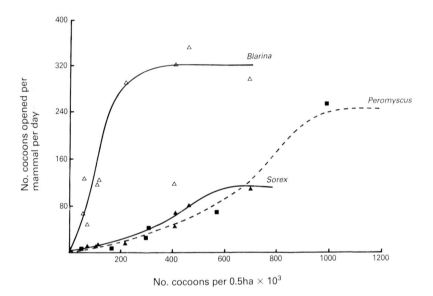

No. cocoons per 0.5ha × 10³

This functional response was not the only observation that Holling made. He also found that certain predator species were apparently responding to increasing availability of prey by increasing their own population numbers—that is they exhibited a numerical response. The response by *Sorex cinereus* was most marked (see Figure 8.2), and both this shrew and the deer mouse showed an initial increase with increasing cocoon density. The short-tailed shrew, which had the lowest population numbers of the three species did not, however, show such a trend.

If both the functional and numerical responses are combined (see Figure 8.3), it can be seen that the total number of prey taken per day by these predators increases rapidly at lower prey densities. But it soon reaches an upper limit and then declines as the predatory impact of the small mammals cannot keep pace with the ever-increasing prey density.

With both functional and numerical responses being exhibited by shrews to differing densities of prey, these predators could have a marked influence on the population dynamics of the European pine sawfly, provided that these insects are not superabundant and that not too many alternative prey are available. Shrews do tend to take a variety of prey when the opportunity arises, and do not concentrate on a single prey type.

Figure 8.2 Numerical response of small mammals to increasing numbers of prey (after Holling, 1959)

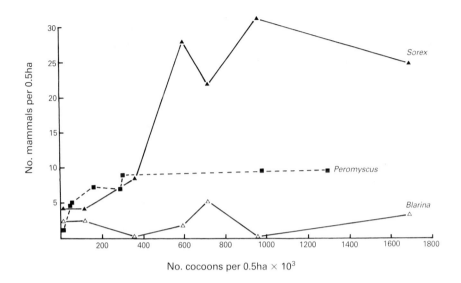

Figure 8.3 Combined functional and numerical responses of small mammals showing their predatory impact on prey populations (after Holling, 1959)

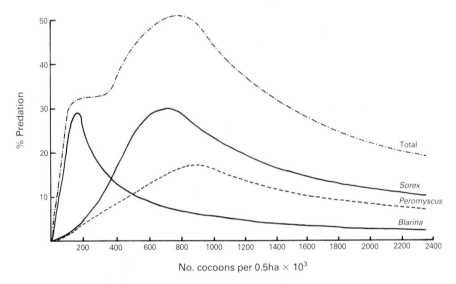

The functional response may be exacerbated by the behaviour of shrews when they discover an excess of prey once their immediate needs have been satisfied, namely their tendency to damage prey without eating them all, and to hoard them. This could increase the effectiveness with which these predators influence the population trends of pest insect populations.

However, numerical responses of mammals to populations of forest insects are seldom observed, and Holling's example appears to be quite exceptional. This may have resulted from the circumstances prevailing: the conifer plantation studied was quite a simple habitat where changes in the abundance of the sawfly cocoons may well have represented a major change in total prey biomass, encouraging shrews to take advantage of increasing numbers of prey by enhancing reproductive output.

Shrews may also have an indirect effect on primary production by eating phytophagous insects, thereby reducing the damage caused by these defoliators. It was estimated by Buckner (1969b) that predation of winter moth (*Operophtera brumata*) by *Sorex araneus* populations could reach 111,150 per hectare per month. The caterpillar of this moth is a major defoliator of oak trees, and periodic outbreaks can cause considerable damage in oak woodlands. Again, it is the larval and pupal stages which are vulnerable to predation by shrews. The larvae descend from the trees in spring to pupate in the soil. They remain as pupae until December when the adults emerge. They are available, therefore, as food when shrew numbers are at their peak, and are widely eaten by them.

Buckner tagged and buried known numbers of winter moth cocoons in the woodland floor in spring and observed their fate a few months later. While some were preyed upon by carnivorous beetles, were parasitised by the wasp *Cyzenis*, or were moribund due to unknown causes, a large proportion (27–65 per cent) were eaten by small mammals, principally shrews.

With their widespread distribution in a range of different habitats and geographical regions, their generalised habits, and their ability to exploit a variety of prey types, shrews might make good agents of biological control to assist in tackling invertebrate pests. This was recognised many years ago in New York State, USA, where efforts were made to protect large plantations from damage by larch sawfly (*Pristiphora erichsonii*) in the mid-1930s by encouraging populations of insectivorous small mammals. Branches pruned from trees were piled on the ground to form hollow domes of brush to provide cover for masked shrews and other small mammals and places for them to nest. Nest boxes were even constructed to encourage populations of deer mice (*Peromyscus maniculatus*), and probably shrews too. These simple techniques proved successful: the numbers of small mammals increased, and there was a reduction in larch sawflies in plantations where they were used compared with plantations where no such encouragement had been provided.

Aware of the potential of shrews as predators of sawfly as demonstrated on the North American mainland, a biological control programme was instigated on the neighbouring island of Newfoundland. As on the mainland, conifer forests in Newfoundland suffer periodic and serious outbreaks of larch sawfly. The small mammal fauna of this island is severely restricted, and shrews are not indigenous. Attempts were made, therefore, to introduce the masked shrew (*Sorex cinereus*), one of the commoner species found on the neighbouring mainland, to augment the natural predators of the pest insect.

Initial attempts at capturing and transporting shrews from the mainland failed dismally since the mortality rate was high. But in September 1958, 22 shrews (twelve females and ten males) were successfully transported by aircraft to Newfoundland where they were promptly released in a bog-type habitat on the west coast of the island. They established themselves there, and at least half of the original number survived through the winter. Breeding commenced in the following spring. By the autumn of 1959, well over 100 shrews occupied the area, including eleven of the original stock and 119 offspring. The shrews continued to increase in number and expand their range. They had dispersal rates of 7–12 miles per year, an impressive feat for such a small mammal, having a mean body weight of a mere 3.6g. Some shrews were relocated to the central part of the island where high populations of larch sawfly were found. By 1977, *S. cinereus* occupied an estimated 80 per cent of the island's 43,000 sq miles.

Follow-up studies revealed that, in areas of high shrew density, the mortality in the larch sawfly population due to predation reached 50 per cent. In areas where sawfly cocoons had relatively high densities but shrew densities were only moderate, predation reached only 10–27 per cent of the sawfly population. This compared with predation rates of sawfly on the mainland in Manitoba of 76–90 per cent in areas of high shrew density and 23–56 per cent in areas of moderate shrew densities. So the level of predation of this pest was somewhat lower in Newfoundland, and the shrews were unlikely to be able to control the sawfly populations. Nevertheless, they did greatly increase the mortality rate of the pest, and contributed more to its demise that any other biological agent present.

SHREWS AS PESTS

Although shrews are primarily predators of invertebrates, they do eat some plant material, and some species seem to take more plant material than others. For example, *Sorex araneus*, whose diet has been studied in detail in different habitats, takes little vegetable matter (although a wild individual has been observed eating porridge oats on a bird table), but *Blarina brevicauda* will readily eat sunflower seeds, beechnuts, Douglas fir seeds and soft corn.

This leads to some speculation about the effects that shrews might have on crop plants. *Sorex trowbridgii* and *S. pacificus yaquinae* also eat Douglas fir seeds. It has been suggested that this habit, coupled with the high population density of these shrews in some areas of the Pacific north-west of the USA, may have a significant effect in checking the regeneration of Douglas fir. However, any such effects are likely to be localised and on a small scale, and there are no reports of shrews causing significant damage to crops. Any harm caused to crop plants, fruits, seeds or stored products are far outweighed by the benefits of their predation on pest insects.

PESTICIDES, POLLUTION AND HABITAT DESTRUCTION

Shrews, being common and ubiquitous, come into contact with many of man's agricultural and industrial activities, and can be useful indicators of pesticides, pollutants and habitat change. Being predators, they accumulate and concentrate chemicals in their bodies by feeding on invertebrates. DDT, for example, is known to accumulate in invertebrates and their predators, and earthworms accumulate large quantities of this toxic chemical, in the range of 1–45 parts per million (Davies, 1971). Earthworms form an important component of the diet of many shrews, including the common shrew (*Sorex araneus*) and the short-tailed shrew *Blarina brevicauda*. Tissues of *B. brevicauda* have been found to contain DDT as long as nine years after an application of this chemical to a forest habitat (Dimond and Sherburne, 1969). Other researchers detected no decline in DDT levels in populations of this shrew even four years after an application of DDT to a grassland habitat.

The predatory habits of shrews, coupled with their high metabolic rates, their rapid throughput of food, and their short life-spans present particular problems with respect to the potential effects of such chemicals and pollutants on these small mammals. What, for example, is the effect of DDT on the physiology of shrews?

An experiment to investigate the effects of DDT on *B. brevicauda* was carried out by Braham and Neal (1974). They took two groups of shrews and fed them on earthworms for a month. The control group of shrews was fed on uncontaminated earthworms, but the experimental group was given earthworms collected from a grassland area which had been treated with DDT four years earlier. Analysis showed that while the uncontaminated earthworms contained no DDT, those from the treated grassland which were fed to the shrews contained 16.6ppm of DDT.

At the conclusion of the experiment, four weeks later, the DDT levels in the liver and fat tissue of both groups of shrews were analysed. The control shrews contained no DDT but the experimental shrews, fed on contaminated earthworms, had accumulated 7.59ppm of DDT in the liver and 14.7ppm in the fat. This clearly demonstrated that accumulation of these pesticides does occur, and quite rapidly.

While not fatal, these levels of DDT in the body did produce certain physiological changes in the shrews. After a week of feeding on DDT-contaminated earthworms, the metabolic rates of the experimental shrews, measured in terms of oxygen consumption, was significantly higher than that of the control shrews. But by the second and third weeks the metabolic rates of the experimental shrews had decreased to their previous level and matched that of the control shrews. So, it seems that exposure to DDT does alter metabolic rates of shrews, but only temporarily.

It was proposed that the initial increase in metabolic rate was due to high levels of DDT circulating in the bloodstream, and that the subsequent decrease was due to the removal of DDT from the blood by liver enzyme activity. When the oxygen consumption was measured after the shrews had been starved for several hours, the metabolic rate again increased, suggesting that DDT was being metabolised along with fat reserves into the bloodstream. So, as fat becomes mobilised under stress conditions, such as a shortage of food or a decrease in ambient temperature, DDT could have an important effect on the survival rate of shrews by increasing the metabolic rate and causing a more rapid use of valuable energy reserves.

Wild mammals and other organisms are exposed to contamination of their habitats by a variety of pollutants produced by man's increasing industrialisation and mechanisation. Contamination of habitats, and the resultant effects on the food chain, by metals released in the mining and refining of metalliferous ores, by motor exhausts, and even by the application of sewage sludge to land, is becoming more widespread. Shrews from contaminated grasslands have been found to accumulate some of the highest concentrations of metals recorded in wildlife. The accumulation of lead, zinc and cadmium by small mammals, including shrews, has been the subject of considerable study, and the nature of the diet is found to have a major influence on the levels of accumulation of different metals.

Cadmium is a toxic by-product of industrial processes which contaminates soils in affected areas. Shrews such as *Sorex araneus* have been found to accumulate higher levels of cadmium than wild voles or mice. This can be explained by the difference in diet between the insectivorous shrews and the herbivorous rodents. Invertebrate prey contain much higher levels of cadmium than vegetation (see Table 8.1). Cadmium becomes chemically transformed in its passage up the food chain and seems to be more readily absorbed through the gut walls of invertebrates and their predators than by plants.

Table 8.1 Accumulation of cadmium by wild shrews and voles
(after Andrews, Johnson and Cooke, 1984)

	Mean concentration of cadmium (μg/g dry weight)			
	Sorex araneus	Inverte- brates	Microtus agrestis	Vegetation
Control site	1.19	2.10	0.88	1.21
Contaminated site	52.70	23.20	1.84	4.75

Old mining sites and spoil heaps are a major source of lead contamination. They soon become colonised by plants and animals, but small mammals from contaminated sites accumulate higher levels of lead than those from uncontaminated control sites. Shrews accumulate lead, as with cadmium, by eating contaminated invertebrates. However, invertebrates do not accumulate lead to nearly the same extent as vegetation. So lead levels in shrews are not as high as those in herbivorous voles from contaminated sites (see Table 8.2). It is thought that the lead ingested with plant material is easily absorbed in the gut, whereas lead consumed with invertebrate tissue is closely bound with the chitinous exoskeleton where it is relatively immobile and less easily digested.

Lead accumulates mostly in the skeleton of mammals, but little is known about the toxicological effects of this metal on small, short-lived animals such as shrews and voles. Cadmium concentrates in the soft tissues, particularly the kidneys, liver and heart. Examination of these target organs by electron microscopy reveals some damage to the tissues in shrews from contaminated sites. This includes lesions in the kidney,

**Table 8.2 Accumulation of lead by wild shrews and voles
(after Roberts, Johnson and Hutton, 1978)**

	Mean concentration of lead (μg/g fresh weight)			
	Sorex araneus	Inverte- brates	Microtus agrestis	Vegetation
Control site	0.9	18.4	2.8	20.8
Contaminated site	11.2	61.9	45.3	120.0

damage to the glomeruli and aberrations and distortions in cell struc- tures such as the mitochondria. Signs of damage and metabolic disrup- tion in the liver include hypertrophy in the epithelium of the sinusoids and unusually high numbers of immature red blood cells in the capillary sinuses suggesting anaemia. However, the effects of this on the mortality rates, survivorship and breeding successes of shrews is not known.

Shrews are adaptable and will establish themselves in all kinds of available habitats, but they require some vegetation cover amongst which to live and hunt for invertebrates. They are common in hedgerows and other marginal habitats including agricultural headlands and areas of waste scrubland, and so are vulnerable to changes in agricultural and industrial practices. Intensive agriculture with its accompanying use of herbicides and pesticides will affect shrews by destroying their habitats and the prey on which they feed. Water shrews are particularly vulner- able to the destruction of their aquatic habitats through pollution and drainage, and *Sorex palustris punctulatus* has been reported to be endangered in the southern Appalachians because of the siltation and pollution of the streams that it inhabits. In Britain, water-cress beds are a favoured habitat for the water shrew *Neomys fodiens*. The modernisa- tion of commercial water-cress beds to permit easier maintenance, including the removal of grass banks and their replacement with con- crete, is causing serious habitat destruction for this shrew, whose population numbers are low and whose occurrence is limited. Other species and subspecies which have very localised distributions may also be endangered. One such species is *Blarina carolinensis shermani*, which has only been found in one small area of south-western Florida where human development is threatening its continued existence.

A number of other shrews are classified as rare and endangered, including *Crocidura odorata goliath* from Cameroon, *C. maquassiensis* from southern Africa, and the Bornean musk shrew (*Suncus ater*), which are known only from a small handful of specimens. But so little is known about many species of shrew throughout the world that their conserva- tion status can only be speculated on. They include such species as Kelaart's long-clawed shrews (*Feroculus feroculus*) found only in a small area of the central highlands of Sri Lanka, and Pearson's long-clawed shrew (*Solisorex pearsoni*) also from Sri Lanka, about which almost nothing is known. Besides, new species are still being described from various parts of the globe.

STUDYING SHREWS

Observing Shrews

While shrews are rarely observed in the wild, they can often be heard, particularly on summer days when they are breeding and the population is at its highest. Shrews are then actively searching for mates, invading each others' home ranges and trying to establish their own ranges as soon as they have vacated the parental nest. This period of activity leads to frequent contacts and altercations between shrews, often manifested in shouting matches which can be heard in hedgerows, grasslands and woods.

It has proved possible to observe wild shrews coming to food put out on bird tables, and even to encourage them to visit specially constructed food tables to which they have access, along with other small mammals. This has been most successful at night, using artificial illumination.

Being so small, shrews leave little evidence of their activities in terms of field signs. They have a plantigrade gait with five toes on each foot, and the typical footprint is shown in Figure 8.4. Small, dark, granular faeces, usually 3–4mm in length, can occasionally be found deposited in runways through the vegetation, those of the water shrews *Neomys fodiens* and *Sorex palustris* being quite distinctive. In these species, the faeces are black and granular in structure, being full of the remains of invertebrate exoskeletons. They are often deposited in middens on the banks of streams, in surface burrows, at burrow entrances, in the lee of rocks at the stream edge, or even sometimes quite prominently on the tops of stones.

Shrews often make small holes for runways through ground vegetation, but these are not very distinctive. However, the water shrew, *Neomys fodiens*, does have quite easily recognisable entrances to its burrows in the banks of streams and ditches. The holes are small and round in cross section (about 2cm in diameter) and, unlike those of bank voles and wood mice, there is little disturbance of the soil or vegetation at the entrance.

Water shrews such as *N. fodiens* leave remains of invertebrate prey, mostly the cases of caddis larvae and shells of molluscs, which can sometimes be found at their feeding sites on the banks of streams.

The presence of crocidurine shrews can be detected, even if they cannot be seen, by their distinctive, slightly sweet, musky odour. The larger, tropical species of *Crocidura* have a very strong scent which is also very persistent, lasting for days or even weeks in vacated nest sites.

Shrews and the Law

Recent changes in legislation in Britain, particularly the introduction of the Wildlife and Countryside Act 1981, have had a significant effect on the protection afforded to a number of mammals, including shrews. Under this Act, all species of shrew in Britain are given partial protection under Schedule 6 which severely restricts methods by which certain mammals, including shrews, dolphins, badgers, wildcats and hedgehogs, may be killed or captured. It is an offence to set traps and snares, or

Figure 8.4 Footprints of the common shrew (Sorex araneus)

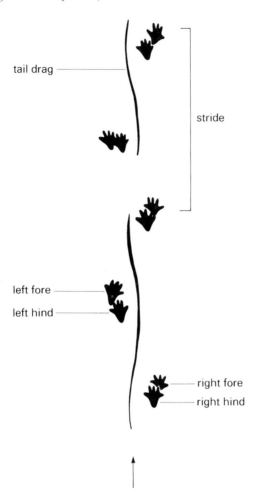

deposit any poisonous or stupefying substances which are calculated to cause bodily injury to these mammals. It is also an offence to use a variety of means for killing or capturing them, such as automatic or semi-automatic weapons, devices for illuminating the target, sound recording used as decoys, and mechanically propelled vehicles for immediate pursuit.

While such restrictions may be more applicable to badgers and dolphins than to shrews, which in any case are generally abundant, the result of the Act is that the trapping (including live-trapping) and killing of shrews is illegal. However, the law states that an offence is not committed if a shrew is caught in a trap or other device which was set in position with the intention of catching unprotected species (such as mice), and that all reasonable precautions were taken to prevent injury to shrews or other Schedule 6 mammals which might accidentally be caught. So when trapping for small mammals other than shrews, some precautions are necessary to minimise the chances of death or damage to shrews. So-called 'reasonable precautions' are subject to individual interpretation, but the recommendation is that when live-trapping is being carried out (for example, with Longworth live-traps) for the purpose of studying rodent populations, shrew holes are provided in the trap. A small hole, with a diameter of 10mm, is cut in the side of the trap, near the base, which will permit any trapped shrew to escape, but not the larger rodents. This, however, is not always a very satisfactory precaution because the water shrew, a relatively large species, cannot escape through such a small hole. Also, young mice and voles, and harvest mice, are small enough to escape by this means, and this may interfere with the objects of the study.

The major cause of death in captured shrews is starvation, since few are able to survive without food for more than three hours. So an alternative precaution is to provide some suitable food, such as blowfly pupae, in the traps in case shrews are captured, and to visit the traps regularly and frequently. At least four visits per 24 hours are recommended, since shrews run out of food rapidly and there are other causes of mortality including stress, cold or overheating. More will be said below about how to catch and keep shrews alive.

If a serious field study of shrews is to be carried out in Britain, this can only be done under licence from the Nature Conservancy Council because of the restrictions imposed by the Wildlife and Countryside Act. In the USA and Canada, wildlife agencies in each state and province have instituted permit procedures to govern scientific collecting of non-game species, including shrews.

Catching and Keeping Shrews

Before the invention of box traps designed to catch small mammals and keep them alive, studies of rodents and shrews were largely restricted to animals caught in break-back mouse traps or pitfall traps. Shrews, with their relatively poor eyesight, their rapid, bustling movements, and their inability to jump great heights, can be captured successfully in pitfall traps set in the ground with their rims flush with the ground surface. Pitfalls can be made from a host of commonly available containers, such as large coffee tins and large plastic yoghurt and ice-cream cartons. For most shrews, pitfalls with a depth of 30cm are sufficient to prevent escape, and a diameter at the rim of about 18cm or more is satisfactory. They are best set in the ground amongst vegetation in the surface runs of small mammals such as voles, where shrews will also be found. If shrews are needed alive, then frequent visits must be made to them, and food must be provided (see below). Waterproof covers mounted on short legs and placed over the pitfalls will keep rainfall out and sunlight off, and so improve survival of shrews. A major drawback of these traps is that they are costly in time and energy to install, large numbers of them are needed for full-scale studies of populations of shrews, and their positioning has to be altered from time to time since the regular pathways used by small mammals change.

The production of specially designed live-traps for small mammals has revolutionised field studies of mice, voles and, of course, shrews. Shrews are immensely curious creatures and readily investigate any new object they encounter in the course of their daily activities. This means that they are easily captured in a range of live-traps, provided the traps can be triggered off by the small weight of the shrew. The most commonly used traps in field studies of small mammals are the Longworth and the Sherman. Both are adjustable in terms of the weight required to trigger them off, but the Longworth is capable of finer adjustment and seems to be more successful at capturing small shrews than the Sherman.

The Longworth trap, designed by Chitty and Kempson in 1949, is probably the most widely used live-trap for small mammals in Britain today. It is illustrated in Figure 8.5. It comprises two sections, a nest box and a tunnel, which are clipped together when in use but are easily disengaged to permit the contents to be extracted when a capture is made. The tunnel section has a hinged door at the front which is held open by a small wire clip attached to a treadle at the base of the back end of the tunnel. When an animal enters the trap and walks over the treadle, the clip holding up the door is moved and the door closes. Bedding material is placed in the nest box, together with a little food, and the captured animal remains secure until it is visited on the next trap round. These traps are suitable for all shrews, but are most efficient at catching the larger species. Although the treadles are adjustable, small shrews which may weigh less than 5g, such as *Sorex minutus*, do not always set off the trap. They are small enough to pass under the treadle without activating the mechanism.

Longworths are made of aluminium, and are relatively light to carry yet hard-wearing. However, they are bulky and heavy to transport in large numbers. Shermans, on the other hand, have the great advantage of being foldable (see Figure 8.6). They are made of finer sheet aluminium and are extremely light. In the USA and Canada, Shermans are the most widely used of the commercially produced live-traps for small mammals. They are also much used for studies in the tropics and sub-tropics. They consist of a single rectangular box with collapsible sides, which allows them to be folded and packed flat. They come in different sizes, and both the small (16 × 5cm) and larger (23 × 7.5cm) ones are suitable for capturing shrews. The door of the trap is held open by a clip and is attached by a spring to a flat panel on the floor of the trap. When the animal enters and walks over the plate, its weight triggers the door mechanism to shut behind it. Although they, too, are adjustable to take account of different body weights of animals to be captured, they are not effective at catching small shrews and, indeed, are much less successful than Longworths. They also have a major drawback in that it is difficult to provide food and bedding inside them without interfering with the trap mechanism which shuts the door. Small shrews can also become wedged under the treadle and die there, undetected.

In order to keep captured shrews alive between visits to traps, food must be provided. For this purpose, blowfly pupae are excellent. They are obtainable all year round by arrangement with local fishing tackle shops which sell bait for anglers. A handful of pupae will keep a shrew going for several hours, and will remain fresh in the traps for several days. Minced beef or other fresh meat is also quite good, but it is messy to use, it is less favoured by the shrews, and in warm weather it goes off quickly and then

Figure 8.5 The Longworth trap: complete and tunnel section showing mechanism (after Gurnell and Flowerdew, 1990)

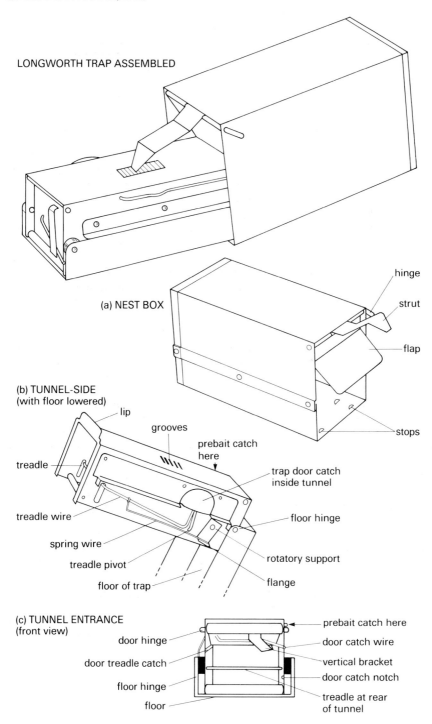

LONGWORTH TRAP ASSEMBLED

(a) NEST BOX

hinge

strut

flap

stops

(b) TUNNEL-SIDE
(with floor lowered)

lip

grooves

prebait catch
here

treadle

trap door catch
inside tunnel

treadle wire

floor hinge

spring wire

treadle pivot

rotatory support

floor of trap

flange

(c) TUNNEL ENTRANCE
(front view)

prebait catch here

door hinge

door catch wire

door treadle catch

vertical bracket

floor hinge

door catch notch

floor

treadle at rear
of tunnel

Figure 8.6 The Sherman trap

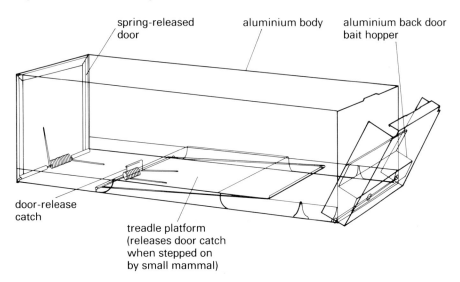

becomes inedible. Other foods such as tinned or dried dog food have similar drawbacks, and are not eaten readily, if at all, by shrews.

The use of suitable bedding material in traps can also minimise mortality. Metal traps such as Longworths are good conductors of heat and will freeze in winter and bake in summer. In winter, it is desirable to use a good insulator such as cotton wool as bedding material, into which shrews can burrow and maintain a warm nest. Hay or straw does not provide sufficient insulation on cold nights, and cotton wool bedding will greatly increase survival. In wet weather, non-absorbent cotton wool is excellent. Although traps such as Longworths are quite waterproof, rain can enter them through the front, and conditions can become damp and cold inside. Once a shrew's coat becomes wet and matted, body heat will be lost rapidly, valuable energy reserves will be used up, and the shrew will quickly die. In hot, sunny weather, careful positioning of the traps amongst the vegetation can reduce mortality from overheating.

None of these precautions can insure against the death of shrews in traps, but they will reduce the chances of mortality. Shrews do not only die of starvation or heat stress in traps, they may also succumb to the stress of being enclosed in a confined space for hours at a time. Shrews do die, even when they have an adequate food supply. So regular and frequent visits to traps are highly desirable. The number and timing of visits depends rather upon weather conditions—more visits are necessary when conditions are very cold or very hot. Generally, four visits per 24 hours is a good, workable regime (dawn, midday, late afternoon and evening). Most shrews will last overnight for up to 10 hours provided they have sufficient food. Pygmy shrews have a lower mortality rate in traps than common shrews, which may be due to their lower absolute food requirements and their ability to tuck their smaller bodies deeper into the nesting material, away from the cold sides of the trap.

A combination of all these precautions should result in very low or no mortality of shrews in traps.

Shrews can be maintained in captivity quite successfully. Most species are not social and so must be kept individually, each in its own cage, or they will fight, often to the death. The exceptions are the crocidurine shrews which can be maintained quite amicably together in small breeding groups. In *Crocidura russula*, for example, a combination of two females and one male works well. Water shrews (*Neomys fodiens*) can also be kept in twos or threes provided there is sufficient space for each to have its own nest site and food source. Although there are frequent bickerings over territorial boundaries, these do not often lead to physical combat.

Shrews are best kept in plastic tanks or wooden boxes which can be easily cleaned. They are great escape artists and can easily get through the wire tops of commercially available mouse and rat cages. Individual shrews require tanks of at least 30 × 15cm, and the larger the better. Small tanks are quickly soiled and so need weekly cleaning. Larger boxes of around 60 × 40cm prove much better since shrews tend to establish defecation sites or middens which can be easily removed without disturbing the whole of the cage and its bedding. Shrews do best with as little disturbance as possible, and so rigorous and frequent cleaning of cages is not recommended. A substratum of 4cm or so of sawdust, peat or soil should be provided, and this should be overlaid with a layer of hay, straw or dry leaf litter. Shrews spend a lot of time exploring, and furrowing through their bedding and the substratum, and this gives them something to occupy their time. Shrews make their own nests and need to have plenty of bedding material, such as hay. They will readily adopt a small cardboard box as a retreat, and build a nest inside it. As this will be kept quite clean, it can remain undisturbed for weeks at a time. If the cage is small enough for the shrew to jump or climb out of, then a top must be fitted, but this should be of wire mesh since perspex causes a build-up of moisture.

One of the greatest problems of maintaining captive shrews is their tendency to become obese. While wild shrews never accumulate fat to any significant extent, many individuals, particularly of *Sorex* species, lay down large deposits of fat around the shoulders after a month or so in captivity. This is presumed to be caused by the relative lack of activity, and abundant but often unsuitable food. This in turn may lead to a premature death. It is difficult to prevent this happening, except by very diligent husbandry. Encouraging activity by providing activity wheels and dispersing food to promote searching behaviour may ease this problem.

Shrews are hardy, and can be kept outdoors all year round provided they have some protection from rain, snow and frost.

Wild shrews have a varied diet comprising a wide range of invertebrates, and this is difficult to simulate in captivity. Shrews can be maintained on substitute diets such as chopped beef, heart and liver, and sometimes tinned dog food, but these foods are messy to use. They also go stale or dry out very quickly, becoming unpalatable to shrews, and so must be provided fresh each day. Shrews readily eat blowfly (*Calliphora* sp) larvae and pupae, and mealworms (larvae of *Tenebrio molitor*), both of which are available commercially. Mealworms, however, are very expensive, and they also tend to cause rapid weight increases since they are rather fatty. Blowfly larvae or pupae are preferable, but need to be supplemented with other foods (such as chopped ox heart and egg) and added nutrients if shrews are to be kept in good condition. Shrews subsisting on pupae alone tend to lose coat condition: the fur becomes rather greasy and matted, and moulting may be incomplete, leaving bald patches. Adding a small quantity of powdered, calcium-based nutrient supplement, available commercially from pet shops, does help. Maggots or pupae can be placed in tall-sided glass pots which allow access by shrews but discourage the prey from escaping. In this way, several days' supply of food can be provided at a time, reducing the amount of maintenance required. Since many shrews show a tendency to hoard food in captivity, regular searching of bedding is desirable to prevent plagues of flies developing.

Shrews do drink and so need access to a source of water. This can be provided in the form of water pots, or as drinking bottles which they quickly learn to use. Water shrews can also be maintained in this way in captivity, and have a greater longevity. It is not necessary to provide open water for them to swim in, but they will readily take to water if it is provided in a container in their cage. The provision of a permanent supply of water for swimming is not recommended in a small cage since the shrews, and the cage, rapidly become sodden.

References

Abe, H. (1982) 'Ecological distribution and faunal structure of small mammals in central Nepal', *Mammalia*, **46**, 477–503

Adams, L.E. (1912) 'Duration of life of the common and lesser shrew, with some notes on their habits', *Mem. Proc. Manchr. Lit. Phil. Soc.*, **56**, 1–10

Allen, J.A. (1917) 'The skeletal characters of *Scutisorex* Thomas', *Bull. Amer. Mus. Nat. Hist.*, **37**, 769–84

Andrews, S.M., Johnson, M.S. and Cooke, J.A. (1984) 'Cadmium in small mammals from grassland established on metalliferous mine waste', *Environ. Poll.*, **33**, 153–62

Ansell, W.F.H. (1964) 'Captivity behaviour and postnatal development of the shrew *Crocidura bicolor*', *Proc. Zool. Soc. Lond.*, **142**, 123–7

Aulak, W. (1967) 'Estimation of small mammal density in three forest biotopes', *Ekologia Polska*, **A15**, 755–78

Banfield, A.W.F. (1974) *The Mammals of Canada*, University of Toronto Press, Toronto

Barnard, C.J. and Brown, C.A.J. (1981) 'Prey size selection and competition in the common shrew (*Sorex araneus*)', *Behav. Ecol. Sociobiol.*, **8**, 239–43

— — (1985a) 'Competition affects risk-sensitivity in foraging shrews (*Sorex araneus*), *Behav. Ecol. Sociobiol.*, **17**, 379–82.

— — (1985b) 'Risk-sensitive foraging in commons shrews (*Sorex araneus*)', *Behav. Ecol. Sociobiol.*, **16**, 161–4

Baxter, R.M. and Meester, J. (1982) 'The captive behaviour of the red musk shrew, *Crocidura flavescens flavescens* (Soricidae: Crocidurinae)', *Mammalia*, **46**, 11–28

Bever, K. (1983) 'Zur nahrung der hausspitzmaus, *Crocidura russula* (Hermann, 1780)', *Säugetier. Mitt.*, **31**, 13–26

Braham, H.W. and Neal, C.M. (1974) 'The effects of DDT on energetics of the short-tailed shrew, *Blarina brevicauda*', *Bull. Environ. Contam. Tox.*, **12**, 32–7

Branis, M. (1981) 'Morphology of the eye of shrews (Soricidae, Insectivora)', *Acta Univ. Carolinae—Biol.*, **11**, 409–45

Brown, J.H. (1975) 'Geographical ecology of desert rodents'. In: *Ecology and Evolution of Communities*, eds Cody, M.L. and Diamond, J.M., Bellknap Press of Harvard University Press, Cambridge, Mass., 315–41

Buchler, E.R. (1976) 'The use of echolocation by the wandering shrew (*Sorex vagrans*)', *Anim. Behav.*, **24**, 858–73

Buckley, J. and Goldsmith, J.G. (1975) 'The prey of the barn owl (*Tyto alba alba*) in east Norfolk', *Mammal Rev.*, **5**, 13–16

Buckner, C.H. (1964) 'Metabolism and feeding behaviour in three species of shrews', *Can. J. Zool.*, **42**, 259–79

167

—— (1966a) 'Populations and ecological relationships of shrews in tamarack bogs of southwestern Manitoba', *J. Mammal.*, **47**, 181–94

—— (1966b) 'The role of vertebrate predators in biological control of forest insects', *Ann. Rev. Ent.*, **11**, 449–70

—— (1969a) 'Some aspects of the population ecology of the common shrew, *Sorex araneus*, near Oxford, England', *J. Mammal.*, **50**, 326–32

—— (1969b) 'The common shrew (*Sorex araneus*) as a predator of the winter moth (*Operophtera brumata*) near Oxford, England', *Can. Entomol.*, **101**, 370–4

Burt, W.H. (1943) 'Territoriality and home range concepts as applied to mammals', *J. Mammal.*, **24**, 346–52

Calder, W.A. (1969) 'Temperature relations and underwater endurance of the smallest homeothermic diver, the water shrew', *Comp. Biochem. Physiol.*, **30**, 1075–82

Chitty, D. and Kempson, D.A. (1949) 'Prebaiting small mammals and a new design of live trap', *Ecology*, **30**, 536–42

Choate, J.R. (1970) 'Systematics and zoogeography of Middle American shrews of the genus *Cryptotis*', *Univ. Kansas Publ. Mus. Nat. Hist.*, **19**, 195–317

Choate, J.R. and Fleharty, E.D. (1973) 'Habitat preference and spatial relations of shrews in a mixed grassland in Kansas', *Southwestern Nat.*, **18**, 110–12

Churchfield, S. (1979) Studies on the ecology and behaviour of British shrews, PhD thesis, University of London

—— (1980a) 'Population dynamics and the seasonal fluctuations in numbers of the common shrew in Britain', *Acta theriol.*, **25**, 415–24

—— (1980b) 'Subterranean foraging and burrowing activity of the common shrew', *Acta theriol.*, **25**, 451–9

—— (1981) 'Water and fat contents of British shrews and their role in the seasonal changes in body weight', *J. Zool., Lond.*, **194**, 165–73

—— (1982a) 'Food availability and the diet of the common shrew, *Sorex araneus*, in Britain', *J. Anim. Ecol.*, **51**, 15–28

—— (1982b) 'The influence of temperature on the activity and food consumption of the common shrew', *Acta theriol.*, **27**, 295–304

—— (1984a) 'An investigation of the population ecology of syntopic shrews inhabiting water-cress beds', *J. Zool., Lond.*, **204**, 229–40

—— (1984b) 'Dietary separation in three species of shrew inhabiting water-cress beds', *J. Zool., Lond.*, **204**, 211–28

—— (1985) 'The feeding ecology of the European water shrew', *Mammal Rev.*, **15**, 13–21

—— (in press) 'Niche dynamics, food resources and feeding strategies in multi-species communities of shrews'. In: *The Biology of the Soricidae, Symposium Vol. Mus. Southwestern Biol.*

Churchfield, S. and Brown, V.K. (1987) 'The trophic impact of small mammals in successional grasslands', *Biol. J. Linn. Soc.*, **31**, 273–90

Corbet, G.B. (1963) 'The frequency of albinism of the tail tip in British mammals', *Proc. Zool. Soc. Lond.*, **140**, 327–30

Corbet, G.B. and Hill, J.E. (1980) *A World List of Mammalian Species*, British Museum (Nat. Hist.) Publ. No. 813

Corbet, G.B. and Southern, H.N. (1977) *The Handbook of British Mammals*. Blackwell Scientific Publications, Oxford

Cowie, R.J. (1977) 'Optimal foraging in great tits (*Parus major*), *Nature*, **268**, 137–9

Crowcroft, W.P. (1954) 'An ecological study of British shrews', D Phil. thesis, University of Oxford

—— (1956) 'On the life span of the common shrew *Sorex araneus*', *Proc. Zool. Soc. Lond.*, **127**, 285–92

—— (1957) *The Life of the Shrew*. Max Reinhardt, London

—— (1964) 'Note on the sexual maturation of shrews (*Sorex araneus* Linnaeus, 1758) in captivity', *Acta theriol.*, **8**, 89–93

Crowcroft, P. and Ingles, J.M. (1959) 'Seasonal changes in the brain-case of the common shrew *Sorex araneus*', *Nature*, **183**, 907–8

Davies, B.N.K. (1971) 'Laboratory studies on the uptake of dieldrin and DDT by earthworms', *Soil. Biol. Biochem.*, **3**, 221–33

Dehnel, A. (1949) 'Studies on the genus *Sorex*', *Ann. Univ. M. Curie-Sklod.*, **4**, 18–102. In Polish

—— (1950) 'Studies on the genus *Neomys*', *Ann. Univ. M. Curie-Sklod.*, **5**, 1–63. In Polish

Delany, M.J. (1964) 'An ecological study of the small mammals in the Queen Elizabeth Park, Uganda', *Rev. Zool. Bot. Afr.*, **70**, 129–49

Dieterlen, F. and Heim de Balsac, H. (1979) 'Zur Ökologie und Taxonomie der Spitzmäuse (Soricidae) des Kivu-gebietes', *Säugetierk. Mitt.*, **27**, 241–87

Dimond, J.B. and Sherburne, J.A. (1969) 'Persistence of DDT in wild populations of small mammals', *Nature*, **221**, 486–7

Dötsch, C. and Koenigswald, W.V. (1978) 'Zur Rotfärbung von Soricidenzähnen', *Z. Säugetierk.*, **43**, 65–70

Eadie, W.R. (1938) 'The dermal glands of shrews', *J. Mammal.*, **19**, 171–4

—— (1952) 'Shrew predation and vole populations on a localised area', *J. Mammal.*, **33**, 185–9

Ellenbroek, F.J.M. (1980) 'Interspecific competition in the shrews *Sorex araneus* and *Sorex minutus* (Soricidae, Insectivora): a population study of the Irish pygmy shrew', *J. Zool., Lond.*, **192**, 119–36

Ewer, R.F. (1968) *The Ethology of Mammals*. Plenum Press, New York

Findley, J.S. (1967) 'Insectivores and dermopterans'. In: *Recent Mammals of the World. A Synopsis of Families*, eds Anderson, S. and Jones, J.K. Jr, Ronald Press, New York, 87–108

Fons, R. (1974) 'Le répertoire comportemental de la pachyure Etrusque, *Suncus etruscus* (Savi, 1822), *Terre Vie*, **1**, 131–57

Ford, C.E., Hamerton, J.L. and Sharman, G.B. (1957) 'Chromosome polymorphism in the common shrew', *Nature*, **180**, 392–3

Forsman, K.A. and Malmquist, M.G. (1988) 'Evidence for echolocation in the common shrew, *Sorex araneus*', *J. Zool., Lond.*, **216**, 655–62

French, T.W. (1984) 'Dietary overlap of *Sorex longirostris* and *S. cinereus* in hardwood floodplain habitats in Vigo County, Indiana', *Am. Midl. Nat.*, **111**, 41–6

Frey, H. (1979) 'La température corporelle de *Suncus etruscus* (Soricidae, Insectivora) au cours de l'activité, du repos normothermique et de la torpeur', *Rev. Suisse Zool.*, **86**, 653–62

Gebczynski, M. (1965) 'Seasonal and age changes in the metabolism and activity of the common shrew *Sorex araneus*', *Acta theriol.*, **22**, 303–31

Genoud, M. (1978) 'Étude d'une population urbaine de musaraignes musettes (*Crocidura russula* Hermann, 1780)', *Bull. Soc. Vaud. Sc. Nat.*, **74**, 24–34

—— (1985) 'Ecological energetics in two European shrews: *Crocidura russula* and *Sorex coronatus* (Soricidae: Mammalia)', *J. Zool., Lond.*, **207**, 63–86

—— (1988) 'Energetic strategies of shrews: ecological constraints and evolutionary implications', *Mammal Rev.*, **18**, 173–93

Genoud, M. and Hausser, J. (1979) 'Ecologie d'une population de *Crocidura russula* en milieu rural montagnard (Insectivora, Soricidae)', *Terre Vie*, **33**, 539–54

George, S.B. (1986) 'Evolution and historical biogeography of soricine shrews', *Systematic Zool.*, **35**, 153–62

Getz, L.L. (1961) 'Factors influencing the local distribution of shrews', *Amer. Midl. Nat.*, **65**, 67–88

Godfrey, G.K. (1978('The breeding season of the lesser white-toothed shrew

(*Crocidura suaveolens* Pallas, 1811) in Jersey', *Bull. Soc. Jersiaise*, **22**, 195–6

—— (1979) 'Gestation period in the common shrew, *Sorex coronatus (araneus) fretalis*', *J. Zool., Lond.*, **189**, 548–51

Gosczynski, J. (1977) 'Connections between predatory birds and mammals and their prey', *Acta theriol.*, **22**, 399–430

Gould, E. (1969) 'Communication in three genera of shrews (Soricidae): *Suncus, Blarina* and *Cryptotis*', *Behav. Biol.*, **A3**, 11–31

Grainger, J.P. and Fairley, J.S. (1978) 'Studies on the biology of the pygmy shrew, *Sorex minutus*, in the west of Ireland', *J. Zool., Lond.*, **186**, 109–41

Grün, G. and Schwammberger, K-H. (1980) 'Ultrastructure of the retina in the shrew (Insectivora: Soricidae), *Z. Säugetierkd.*, **45**, 207–16

Gureev, A.A. (1971) *Shrew (Soricidae) fauna of the World*. Acad. Sci., USSR, Zool. Inst., Nauka Publishing House, Leningrad. In Russian

Gurnell, J. and Flowerdew, J.R. (1990) *Live Trapping Small Mammals. A Practical Guide*. Mammal Society Occasional Publication, second edn

Hamilton, W.J. (1930) 'The food of the Soricidae', *J. Mammal.*, **11**, 26–39

Harper, R.J. (1977) ' "Caravanning" in *Sorex* species', *J. Zool., Lond.*, **183**, 541

Hawes, M.L. (1976) 'Odor as a possible isolating mechanism in sympatric species of shrews (*Sorex vagrans* and *S. obscurus*)', *J. Mammal.*, **57**, 404–6

—— (1977) 'Home range, territoriality and ecological separation in sympatric shrews, *Sorex vagrans* and *Sorex obscurus*', *J. Mammal.*, **58**, 354–67

Hawkins, A.E. and Jewell, P.A. (1962) 'Food consumption and energy requirements of captive British shrews and the mole', *Proc. Zool. Soc. Lond.*, **138** 137–55

Heikura, K. (1984) 'The population dynamics and the influence of winter on the common shrew (*Sorex araneus* L.)'. In: *Winter Ecology of Small Mammals*. ed. J.F. Merritt. Special Publication Carnegie Museum Nat. Hist., **10**, 343–61

Hoffmeister, D. F. and Goodpaster, W.W. (1962) 'Life history of the desert shrew *Notiosorex crawfordi*', *Southwestern Nat.*, **7**, 236–52

Holling, C.S. (1955) 'The selection by certain small mammals of dead, parasitised and healthy prepupae of the European pine sawfly, *Neodiprion sertifer* (Geoff)', *Can. J. Zool.*, **33**, 404–19

—— (1958) 'Sensory stimuli involved in the location and selection of sawfly cocoons by small mammals', *Can. J. Zool.*, **36**, 633–53

—— (1959) 'The components of predation as revealed by a study of the small mammal predation of European sawfly', *Can. Entomol.*, **91** 293–332

Hutterer, R. (1976) 'Beobachtungen zur Geburt und Jugendentwicklung der Zwergspitzmaus, *Sorex minutus* L. (Soricidae: Insectivora)', *Z. Säugetierkd.*, **41**, 1–22

—— (1977) 'Haltung und lebensdauer von Spitzmäusen der Gattung *Sorex* (Mammalia, Insectivora)', *Z. angewandte Zool.*, **64**, 353–67

—— (1978) 'Courtship calls of the water-shrew (*Neomys fodiens*) and related vocalisations of further species of Soricidae', *Z. Säugetierkd.*, **43**, 330–6

—— (1985) 'Anatomical adaptations of shrews', *Mammal Rev.*, **15**, 43–55

Hutterer, R. and Hürter, T. (1981) 'Adapative haarstrukturen bei wasserspitmäusen (Insectivora, Soricinae)', *Z. Säugetierkd.*, **46**, 1–11

Hyvarinen, H. (1967) 'Variation of the size and cell types of the anterior lobe of the pituitary during the life cycle of the common shrew (*Sorex araneus* L.)', *Aquilo Series Zoologica*, **5**, 35–40

—— (1969) 'On the seasonal changes in the skeleton of the common shrew (*Sorex araneus* L.) and their physiological background', *Aquilo Series Zoologica*, **7**, 1–32

Ingram, W.M. (1942) 'Snail associates of *Blarina brevicauda talpoides* (Say)', *J. Mammal.*, **23**, 255–8

Jeanmaire-Bescançon, F. (1986) 'Estimation de l'âge et de la longévité chez *Crocidura russula* (Insectivora: Soricidae)', *Acta Oecol, Oecol. Appl.*, **7**, 355–66

Jones, C.A., Choate, J.R. and Genoways, H.H. (1984) 'Phylogeny and paleo-biogeography of short-tailed shrews (genus *Blarina*)'. In: *Contributions in Quaternary vertebrate paleontology: a volume in memorial to John E. Guilday*, eds Genoways, H.H. and Dawson, M.R., Carnegie Mus. Nat. Hist. Spec. Publ., **8**, 1–538

Kale, H.W. (1972) 'A high concentration of *Cryptotis parva* in a forest in Florida', *J. Mammal.*, **53**, 216–18

Kangur, R. (1954) 'Shrews as tree seed eaters in the Douglas fir region', *Ore. St. Bd. For. Salem Res.*, Note 17

Kirkland, G.L. Jr (In press) 'Community structure and population dynamics'. In: *The Biology of the Soricidae, Symposium Vol. Mus. Southwestern Biol.*

Kisielewska, K. (1963) 'Food composition and reproduction of *Sorex araneus* in the light of parasitological research', *Acta theriol.*, **7**, 127–53

Kleiber, M. (1961) *The Fire of Life*. John Wiley, New York

Krebs, J.R., Erichsen, J.T., Webber, M.I. and Charnov, E.L. (1977) 'Optimal prey selection in the great tit (*Parus major*)', *Anim. Behav.*, **25**, 30–8

Krebs, J.R., Kacelnik, A. and Taylor, P.J. (1978) 'Optimal sampling by foraging birds: an experiment with great tits (*Parus major*)', *Nature*, **275**, 27–31

Lardet, J.–P. (1988) 'Evolution de la température corporelle de la musaraigne aquatique (*Neomys fodiens*) dans l'eau', *Rev. Suisse Zool.*, **95**, 129–35

Lavrov, N.F. (1943) 'Contributions to the biology of the common shrew (*Sorex araneus* L.)', *Zool. Zh.*, **22**, 361–64. In Russian

Lewis, J.W. (1968) 'Studies on the helminth parasites of voles and shrews from Wales', *J. Zool., Lond.*, **154**, 313–31

Lynch, C.D. (1983) 'The mammals of the Orange Free State', *Mem. Nas. Mus. Bloemfontein*, **18**, 20–32

MacArthur, R.H. (1965) 'Patterns of species diversity', *Biological Review*, **40**, 510–33

—— (1972) *Geographic Ecology*. Harper and Row, New York

MacArthur, R.H. and Wilson, E.O. (1967) *The Theory of Island Biogeography*. Princeton University Press, Princeton

Malmquist, M.G. (1985) 'Character displacement and biogeography of the pygmy shrew in northern Europe', *Ecology*, **66**, 372–7

Martin, I.G. (1981) 'Venom of the short-tailed shrew (*Blarina brevicauda*) as an insect immobilising agent', *J. Mammal.*, **62**, 189–92

—— (1984) 'Factors affecting food hoarding in the short-tailed shrew *Blarina brevicauda*', *Mammalia*, **48**, 65–71

McNab, B.K. (1983) 'Energetics, body size, and the limits to endothermy', *J. Zool., Lond.*, **199**, 1–29

Medway, Lord (1978) *The Wild Mammals of Malaya (Peninsula Malaysia) and Singapore*. Oxford University Press, Kuala Lumpur

Merritt, J. (1986) 'Winter survival adaptations of the short-tailed shrew (*Blarina brevicauda*) in an Appalachian montane forest (USA)', *J. Mammal.*, **67**, 450–64

Meylan, A. (1964) 'Le polymorphisme chromosomique de *Sorex araneus* L. (Mammalia—Insectivora)', *Rev. Suisse Zool.*, 371, 903–83

Meylan, A. and Hausser, J. (1978) 'Le type chromosomique A des *Sorex* du group *araneus*: *Sorex coronatus* Millet, 1828 (Mammalia, Insectivora)', *Mammalia*, **42**, 115–22

Michielsen, N. Croin (1966) 'Intraspecific and interspecific competition in the shrews *Sorex araneus* L. and *Sorex minutus* L.', *Arch. Néerlandaises de Zool.*, **17**, 73–174

Mock, O.B. and Conaway, C.H. (1975) 'Reproduction of the least shrew (*Cryptotis parva*) in captivity'. In: *The Laboratory Animal in the study of Reproduction*, eds Antikatzides, T., Erichsen, S. and Spiegel, A., Gustav Fischer Verlag, Stuttgart

Moore, A.W. (1942) 'Shrews as a check on Douglas fir regeneration', *J. Mammal.*, **23**, 37–41

Moors, P.J. (1975) 'The food of weasels (*Mustela nivalis*) on farmland in north-east Scotland', *J. Zool., Lond.*, **177**, 455–61

Moraleva, N.V. (1989) 'Intraspecific interactions in the common shrew *Sorex araneus* in central Siberia', *Ann. Zool. Fennici*, **26**, 4, 425–32

Myrcha, A. (1969) 'Seasonal changes in caloric value, body water and fat in some shrews', *Acta theriol.*, **14**, 211–27

Neet, C.R. and Hausser, J. (1990) 'Habitat selection in zones of parapatric contact between the common shrew *Sorex araneus* and Millet's shrew *S. coronatus*', *J. Anim. Ecol.*, **59**, 235–50

Niethammer, J. (1956) 'Das Gewicht der Waldspitzmaus, *Sorex araneus* Linné, 1758, in Jahreslauf', *Säugetierk. Mitt.*, **4**, 160–5

Novak, R.M. and Paradiso, J.L. (1983) *Walker's Mammals of the World*. Johns Hopkins University Press, Baltimore

Pasanen, S. and Hyvarinen, H. (1970) 'Seasonal variation in the activity of phosphorylase in the interscapular brown fat of small mammals active in winter', *Aquilo Series Zoologica*, **10**, 37–42

Pattie, D. (1973) '*Sorex bendirii*', *Mammalian Species No. 27*. American Society of Mammalogists

Pearson, O.P. (1944) 'Reproduction in the shrew *Blarina brevicauda*', *Amer. J. Anat.*, **75**, 39

—— (1946) 'Scent glands of the short-tailed shrew', *Anat. Rec.*, **94**, 615–29

Pearson O.P. and Pearson A.K. (1947) 'Owl predation in Pennsylvania with notes on the small mammals of Delaware County', *J. Mammal.*, **28**, 137–47

Pernetta, J.C. (1976) 'Diets of the shrews *Sorex araneus* L. and *Sorex minutus* L. in Wytham grassland', *J. Anim. Ecol.*, **45**, 899–912

—— (1977a) 'Population ecology of British shrews in grassland', *Acta theriol.*, **22**, 279–96

—— (1977b) 'Anatomical and behavioural specialisations of shrews in relation to their diet', *Can. J. Zool.*, **55**, 1442–53

Pianka, E.R. (1971) 'Species diversity'. In: *Topics in the Study of Life: the Biosource Book*. Harper and Rowe, New York

—— (1973) 'The structure of lizard communities', *Ann. Rev. Ecol. System.*, **4**, 53–74

Platt, W.J. (1976) 'The social organisation and territoriality of short-tailed shrew (*Blarina brevicauda*) populations in old-field habitats', *Anim. Behav.*, **24**, 305–18

Platt, W.J. and Blakley, N.R. (1973) 'Short-term effects of shrew predation upon invertebrate prey sets in prairie ecosystems', *Proc. Iowa Acad. Sci.*, **80**, 60–6

Price, M. (1953) 'The reproductive cycle of the water shrew *Neomys fodiens bicolor*', *Proc. Zool. Soc. Lond.*, **123**, 599–621

Pucek, M. (1959) 'The effect of the venom of the water shrew *Neomys fodiens* on certain experimental animals', *Acta theriol.*, **3**, 93–104

—— (1967) 'Chemistry and pharmacology of insectivore venoms'. In: *Venomous Animals and their Venoms*, eds Bucherl, W., De..lofen V. and Buckley, E.F. Academic Press, New York, 43–50

—— (1969) '*Neomys anomalus* Cabrera, 1907—a venomous mammal', *Bull de L'Acad. Polon. des Sci.*, **2**, 569–73

Pucek, Z. (1965) 'Seasonal and age changes in the weight of internal organs of shrews', *Acta theriol.*, **10** 369–438

—— (1970) 'Seasonal and age changes in shrews as an adaptive process', *Sym. Zool. Soc. Lond.*, **26**, 189–207

Randolph, J. Collier (1973) 'Ecological energetics of a homeothermic predator, the short-tailed shrew', *Ecology*, **54**, 1166–87

Randolph, S.E. (1975) 'Seasonal dynamics of a host-parasite system: *Ixodes trianguliceps* (Acarina: Ixodidae) and its small mammal hosts', *J. Anim. Ecol.*, **44**, 425–49

Repenning, C.A. (1967) 'Subfamilies and genera of the Soricidae', *U.S. Geol. Surv. Prof. Pap.*, **565**, 1–69

Reumer, J.W.F. (1984) 'Ruscinian and early Pleistocene Soricidae (Insectivora, Mammalia) from Tegelen (The Netherlands) and Hungary', *Scripta Geol.*, **73**, 1–73

—— (1987) 'Redefinition of the Soricidae and the Heterosoricidae (Insectivora, Mammalia), with the description of the Crocidosoricinae, a new subfamily of Soricidae', *Revue de Paléobiologie*, **6**, 189–92

Roberts, R.D., Johnson, M.S. and Hutton, M. (1978) 'Lead contamination of small mammals from metalliferous mines', *Environ. Poll.*, **15**, 61–8

Robinson, D.E. and Brodie, E.D. (1982) 'Food hoarding behaviour in the short-tailed shrew *Blarina brevicauda*', *Amer. Midl. Nat.*, **108**, 369–75

Rood, J.P. (1958) 'Habits of the short-tailed shrew in captivity', *J. Mammal.*, **39**, 499–507

—— (1965) 'Observations on population structure, reproduction and moult of the Scilly shrew', *J. Mammal.*, **46**, 426–33

Rowe-Rowe, D.T. and Meester, J. (1982) 'Population dynamics of small mammals in the Drakensberg of Natal, South Africa', *Z. Säugetierkd.*, **47**, 347–56

Rudd, R.L. (1955) 'Age, sex and weight comparisons in three species of shrews', *J. Mammal.*, **36**, 323–39

Rudge, M.R. (1965) *Feeding and metabolism of the common shrew*. PhD thesis, University of Exeter

—— (1968) 'Food of the common shrew *Sorex araneus* in Britain', *J. Anim. Ecol.*, **37**, 565–81

Savage, D.E. and Russell, D.E. (1983) *Mammalian Paleofaunas of the World*. Addison-Wesley, Reading, Mass.

Schlueter, A. (1980) 'Forest shrews (*Sorex araneus*) and water shrews (*Neomys fodiens*) as carrion eaters in winter', *Säugetierk. Mitt.*, **28**, 45–54

Schmidt, U. (1979) 'Die lokalisation vergrabenen Futters bei der Hausspitzmaus, *Crocidura russula* Hermann', *Z. Säugetierkd.*, **44**, 59–60

Searle, J.B. (1986) 'Factors responsible for a karyotypic polymorphism in the common shrew *Sorex araneus*', *Proc. Roy. Soc. Lond.*, **B229**, 277–98

Searle, J.B. and Wilkinson, P.J. (1987) 'Karyotypic variation in the common shrew (*Sorex araneus*) in Britain—a "Celtic Fringe" ', *Heredity*, **59**, 345–51

Sheppe, W.A. (1973) 'Notes on Zambian rodents and shrews', *Puku*, **7**, 167–90

Shillito, J.F. (1960) 'The general ecology of the common shrew *Sorex araneus* L.', PhD thesis, University of Exeter

—— (1963) 'Field observations on growth, reproduction and activity of a woodland population of the common shrew *Sorex araneus*', *Proc. Zoo. Soc. Lond.*, **140**, 99–114

Skarén, U. (1964) 'Variation in two shrews, *Sorex unguiculatus* Dobson and *S. araneus* L.', *Ann. Zool. Fennici*, **1**, 94–124

—— (1973) 'Spring moult and onset of the breeding season of the common shrew *Sorex araneus* in central Finland;, *Acta theriol.*, **18**, 443–58

Smith, J.N.M. and Sweatman, H.P.A. (1974) 'Food searching behaviour of titmice in patchy environments', *Ecology*, **55** 1216–32

Sorensen, M.W. (1962) 'Some aspects of water shrew behaviour', *Amer. Midl. Nat.*, **68**, 445–62

Southern, H.N. (1954) 'Tawny owls and their prey', *Ibis*, **96**, 384–408

Spencer, A.W. and Pettus, D. (1966) 'Habitat preferences in five sympatric species of long-tailed shrews', *Ecology*, **47**, 677–83

Stein, G.H.W. (1950) 'Zur Biologie des Maulwurfs, *Talpa europaea* L.', *Bonn. Zool. Beitr.*, **1**, 97–116

Tapper, S.C. (1976) 'The diet of weasels, *Mustela nivalis* and stoats *Mustela erminea* during early summer, in relation to predation on gamebirds', *J. Zool., Lond.*, **179**, 219–24

Terry, C.J. (1981) 'Habitat differentiation among three species of *Sorex* and *Neurotrichus gibbsi* in Washington', *Amer. Midl. Nat.*, **106**, 119–25

Thorpe, J.P. (1982) 'The molecular clock hypothesis: biochemical evolution, genetic differentiation and systematics', *Ann. Rev. Ecol. System.*, **13**, 139–68

Tomasi, T.E. (1978) 'The function of venom in the short-tailed shrew, *Blarina brevicauda*', *J. Mammal.*, **59**, 852–4

—— (1979) 'Echolocation by the short-tailed shrew *Blarina brevicauda*', *J. Mammal.*, **60**, 751–9

Vlasak, P. (1970) 'The biology of reproduction and postnatal development of *Crocidura suaveolens* Pallas, 1811 under laboratory conditions', *Acta Univ. Carolinae—Biol.*, 207–92

Vogel, P. (1972) 'Beitrag zur Fortpflanzungsbiologie der Gattungen *Sorex, Neomys* und *Crocidura* (Soricidae)', *Verhandl. Naturf. Ges. Basel.*, **82**, 165–92

—— (1974) 'Note sur le comportement arboricole de *Sylvisorex megalura* (Soricidae, Insectivora)', *Mammalia*, **38**, 171–6

—— (1976) 'Energy consumption of European and African shrews', *Acta theriol.*, **21**, 195–206

—— (1990) 'Body temperature and fur quality in swimming water shrews, *Neomys fodiens* (Mammalia, Insectivora)', *Z. Säugetierkd*, **55**, 2, 73–80

Wołk, K. (1976) 'The winter food of the European water shrew', *Acta theriol.*, **21**, 117–29

Wrigley, R.E., Dubois, J.E. and Copland, H.W.R. (1979) 'Habitat, abundance and distribution of six species of shrews in Manitoba, Canada', *J. Mammal.*, **60**, 505–20

Yalden, D.W. (1981) 'The occurrence of the pygmy shrew, *Sorex minutus*, on moorland and the implications for its presence in Ireland', *J. Zool., Lond.*, **195**, 147–56

Yalden, D.W., Morris, P.A. and Harper, J. (1973) 'Studies on the comparative ecology of some French small mammals', *Mammalia*, **37**, 257–76

Yates, T.L. (1984) 'Insectivores, elephant shrews, tree shrews and dermopterans'. In: *Orders and Families of Recent Mammals of the World*, Second Edition, eds Anderson, S. and Jones, J.K. Jr, John Wiley, New York, 117–44

Yoshino, H. and Abe, H. (1984) 'Comparative study on the foraging habits of two species of soricine shrews', *Acta theriol.*, **29**, 35–43

Yudin, B.S. (1962) 'Ecology of shrews (genus *Sorex*) of western Siberia, USSR'. In: *Problems on the Ecology, Zoogeography and Systematics of Animals*, **8**, 33–134. In Russian

Yudin, B.S., Galkina, L.I. and Potarkina, A.F. (1979) *Mammals of the Altai-Sayan Mountain Systems*. Nauka Siberian otd., Novosibirsk. In Russian

Index

General Index

Index to Species and Genera of Shrews